AUTISM

LIFE IN THE PRISM

KRISTINA DESJARDINS

AAPC PUBLISHING
PO Box 861116
Shawnee, KS 66286
Local Phone (913) 897-1004 Fax (913) 728-6090
www.aapcautismbooks.com

Printed in the United States of America

Published December 2021 by AAPC Publishing

Names: DesJardins, Kristina, author.
Title: Autism : life in the prism / Kristina DesJardins.
Description: Shawnee, KS : AAPC Publishing, [2021]
Identifiers: ISBN: 978-1-956110-06-7
Subjects: LCSH: Autistic people--Life skills guides. | Autistic youth--Life skills guides. | Autistic children--Life skills guides. | Parents of autistic children--Handbooks, manuals, etc. | Teachers of children with disabilities--Handbooks, manuals, etc. | LCGFT: Handbooks and manuals.
Classification: LCC: RC553.A88 D47 2021 | DDC: 616.85/882--dc23

CONTENTS

FOREWORD

This is Kris. I am the author of this book and the creator and editor of the website chat1autism.weebly.com. You will find that a little of the information given in this book is also in the website since not everyone can afford to buy my book. Although a little of the information here is also in my website, my personal experiences and expanded explanations, more tips, etc., is included to make this a very valuable resource. When you do see the words in this book like (*check my website…*), it means that although the information will not be in this book, it will be a notation on my website, so be sure to check it for the references to make this particular experience a digital as well. Digital, meaning videos and pictures to help you understand *visually* my written explanations, tips, etc., that I have in this book. (*In addition, the same vice versa*) I have been told by those who have read my book, "I found this book to be extremely beneficial because of all the added context, the in-depth explanations, topics, personal information, etc., that is not on your website. It is wonderful and I think an amazing value to get digital extras like graphics, your reference links, videos, your posted studies on the web pages, slideshows, reports, the polls/questionnaire page, threads and chat rooms, and an autism movie trailer page along with this book that brings so much more understanding through your precise and unique writing."

I do hope that my large website will still be something you can use upon completion of reading my book, for those extra digital favors. With eras and upgrades, certain websites or companies may change, or videos, so I will be updating these things as needed. I have added a separate guestbook for those who have read, *Autism: Life in the Prism.*

I also want you to know that in one of my specific chapters, chapter 11 on personal questions, I answer questions that could very well be answered in the other chapters to do with the specific subjects, but I thought they should be spread out. On social media you can contact me with my full name via my Facebook or on Twitter at KEDesJardins. My email is also a great way to get in touch with me. I really with all my heart hope you enjoy and take the time to soak everything in that I have written as I really want to bring up issues as well, that for whatever reasons are not brought up much in the topic of autism, and that are so very important.

1

A SUMMARY OF MY EARLY/YOUNGER LIFE AND THE SIGNS OF AUTISM THAT WENT UNNOTICED IN THE BEGINNING

Life started out as any life would hope to be. "It's a girl!" That was the sign my dad constructed and hung above our garage door for the entire neighborhood to stop and see. I was born 6 pounds, 8 ounces, on Aug. 6, 1986. The morning sun was shining down upon the state of 10,000 lakes. Each reflection to my parents was my warm eyes staring back at them. The nurses smeared black ink all over my tiny hands and feet. How perfect they looked. My brothers, anxiously awaiting by my mom's hospital bed (*after the delivery and clean up*), their new baby sister. I did not even get to feel the emotions since I was only minutes old, to come to this great big place with noises, faces, lights, colors and smells. What a different place than before!

The doctors had originally told my parents, "You are going to have a large baby," but there I came out, very tiny and scrawny. The doctors were surprised. The pupils of my eyes were very large, almost dilated like. This can be a physical sign of autism. A psychologist brought this physical symptom to both my parent's attention during my diagnosis, many years later.

Days later, my home, my room freshly painted and decorated, welcomed me with an oak crib covered in yellow and white plaid. A rainbow hung slightly above my head, hovering the crib, while a silver cross representing Christ dangled by a wire on the wall, just in the middle of it. Since the day I was born, I would sleep all through the night for the first few years, unusual to other babies. I would scream, cry, vomit and screech all day long. Constant humming sounds, rocking and swinging, bouncing and low lights would calm me. My feedings were unusual, as I did not want to eat anything. When I did, I would take very little. Colic and a non-desire, sometimes a strong desire not to eat, is a trait of autism. My parents did not know this and the doctors never thought about it. Stomach problems acquired and I would get sick after eating.

I do not remember a great very many people holding me in their arms when I was home for those first 12 months, but do remember one very clearly. My great grandmother of my father sat in a nursing home and held me in her arms before leaving her body. Her smile and hair stick in my mind even though I was only a year old. I guess true cherished moments are shared with everyone, even the smallest of them. This ability to remember, especially from such young years, is acknowledgeable by psychologists as a trait of savantism in autism.

When I was about a three and a half years old, my mom kissed me extra-long and told me she would be back in a week. (*I of course did not understand what she meant*) Each night she was gone, I remember laying my head on my grandmother's bear to listen to its built in heartbeat. Finally one morning, waking up groggy as a young child does, I remember seeing my mom at the top of the stairs leading to the basement. Something colorful with patterns lay gently over her shoulder. In addition, she carried many books (*those things she read to me*) and they had silver binding on the edges that faced me. Undoubtedly, I think I was more interested in that rather than the book itself, as I would spend long amounts of time staring and examining the silver binding. Being more interested in patterns rather than socially interacting or listening to a story is another symptom of autism. My eyes had looked somewhat glossy as I stared at them and smiled. From the small age of two years, I crawled to our fireplace area. A gold and shiny cradle rocked by the fireplace. Lots of flaking wood filled its bed. I told my parents my memory of this when I was 17 and they were amazed as well at my recall of several family events, where we sat in restaurants, family conversations I could repeat back, etc. I gently and carefully took each stick of wood and laid it gently on the carpet. The bed, dipped like a Pringle, was the perfect shape for a tiny body. I crawled under the handle and laid my body onto the cool metal. I grabbed the strands of carpet with my fingers to make it rock. I would lay there for however long I could. I felt so safe and quiet. Similarly, I would lie in my toy box to get the same feeling. I did not rely on anything else but the wood holder that I turned into my rocking sanctuary! "Oh, how cute!" everyone there would say. Our sweet Labrador retriever would run up and lay with me beside the wood cradle holder.

However, what my family did not know, was *why* I would spend long times laying and rocking in the cradle that was meant for wood. Moreover, they never thought it over until quite a long time had passed, finally questioning it years later. Rocking, being alone and enjoying being curled up is also a red flag of autism that was not noticed by my parents or other family members.

I also used to ask my brothers to sit on the cushions I laid on top of me. I also asked them to make a fort made out of the cushions and pick me up and throw me into the fort. I wanted to feel my body and wanted to get that strong input. We did this as a routine, and it had to be done very specifically. I would take two bites of toast, and then I would get picked up and tossed into the fort. They would rebuild the fort of cushions while I took another two bites of the toast, and then they would do it again. This neediness for routine is also another red flag for autism. It had to be done in that order or I would get upset. I loved the feeling of being tossed into cushions as well as my cousin dropping heavily weighted toys onto me while I would lie down. This need for deep pressure input was also a very big flag for autism, especially the constant need for it.

I remember my brothers used to hold me and talk about their 'little sister' to all their friends. Moreover, like any other two (or more) brothers, they had birthdays. The parties were fun for them with lots of kids playing together, but I cried and screamed. When I would finally calm down, I would walk around staring at the ceiling and lights. I would bump into my brothers and their friends while they were playing "pin the tail on the donkey." My brothers would pick me up and move me out of the way, and I rather screeched while they did. Only lots of bouncing and moving me to my own area of the house where the people, lights, noises and movement was gone could calm me down. I would make what others would call "funny noises," while all the time it was supposed to be identified as vocal stimming. Even at my own birthdays, I would scream and cry when other people would sing to me or be around me. This is overstimulation. Before the people would get there to celebrate, my brothers laughed over how I was 'throwing myself to the ground over and over." My brothers would pick me up each time, laugh and tell everyone to watch. They all thought it was a cute quark. Moreover, once again, it was a red flag of autism. I was gathering input that was lacking in my system.

I had opened one of my presents when I was about three years old. The present was a pretend vacuum that lit up and made a suction sound. I kept vacuuming everything I could and would not stop to open any of my other presents. After about 15 minutes, my brothers and parents tried re-directing me to the other presents I had not yet opened. I started crying and saying, "Vac, vac…" Still to this day, I love vacuums (*the ones that do not hurt my ears*) and even asked for a sweeper one for Christmas just to have. When in the store, I would be stuck in the vacuum section. This also is a more-than-not common trait (*along with trains, laces, etc.*) when it comes to obsessions and autism.

My mother worked as a daycare assistant at an available room in a civic center. She would take me along with her, as I was not old enough or mature enough to enter school. I would lie on the carpet and rub it constantly for an incredibly long amount of time. It felt so good to my hands, and I could then, feel my body. I would make lines by rubbing the carpet strands upward. This, as they had not noticed, was another symptom of autism known as tactile self-stimulation. Still to this day, sometimes if you come to my house, you will see an entire section in a room with carpet strands turned up into lines because I had been doing that for an incredibly large amount of time. I used to do the same thing with a deck of cards and start from my room, down the stairs and all around the house!

Many people have said to me, "My child has spit all over her clothes, her bedding and all over her, yet she doesn't seem to mind it. Why is this and how do I stop it?" Along with rubbing the carpet, I used to purposely salivate on my bed sheets, floors or carpets so I could feel the texture with my hands and face. Still to this day, my bed sheets or clothes seem to appear wet at times, as I am unconsciously doing this. When I am done, I realize what I have done. This is also a form of tactile stimulation both in the outer body and in the mouth. The salivating a lot also happened in the center, but the teachers and my parents were so busy they did not notice it happening.

In the civic center, there was one toy in particular. It was a clear, funnel shaped toy with a yellow button on top of it. You would press the button down and the fan would spin to create a picture of an orange and purple butterfly with pink in it. (*The carpet and especially the specifics of the toy is another representation of memorization categorized as savant abilities w/in the autism spectrum*) I would do it repeatedly for half hours at a time. I remember getting the same lipstick candy at the end of the day that was served at the snack table across from the partially enclosed pool with winding blue and white slides. The smell is still very distinct, a smell of which I love… Chlorine. I would sit by the dishwasher at home and with both my eyes closed, would put my face near the top of the opening so I could smell the scent. This is self-olfactory stimulation.

My brothers used to tell my mother that they heard me talking to myself in the bathroom or in the house. This is also a symptom of autism. As far as that was concerned, I would also spend more time in the bathroom because I was wrapping myself up in the shower curtain. It made me feel a lot better. I would also hide in the closet behind the clothes, where it was dark and I could be by myself in my own little box. Similarly, I would again, find another place to hide; an empty toy box with sliding top doors meant to hold board games.

Similar in nature to the constant fascination and play with the funnel toy, I would spend large amounts of time spinning a ball with beads in it on a hand toy. My parents would call my name and I would not hear them or respond even to touch. In a video that my parents had made, my eyes got large and it sounded like I was sucking spit in my mouth because I was so excited and intrigued in the spinning. My parents found it cute, once again, not knowing that it was a red flag of autism. This also went for the wheels on cars, the spinning of bike tires, fans, etc. I used to love to watch the cars and marbles go down a racetrack that my grandmother had in her basement. I would love the way it looked going down the slides and turns, etc. It was visually exciting and calming at the same time. I also loved to carry around a lace, like the ones that would be in your shoes. It was very important to me and I used to love looking at it, especially in school. I would line them up and watch things at school spin when there was free time, either at my desk or in the back corner of the room. This should have been reported to the school psychologist, as all the symptoms of autism were there. The obsessions with linear objects such as the laces, being alone in a corner during playtime, and lining things up are classic symptoms of autism. I am going to now include a picture of what I would do some of the time with cards and pens below.

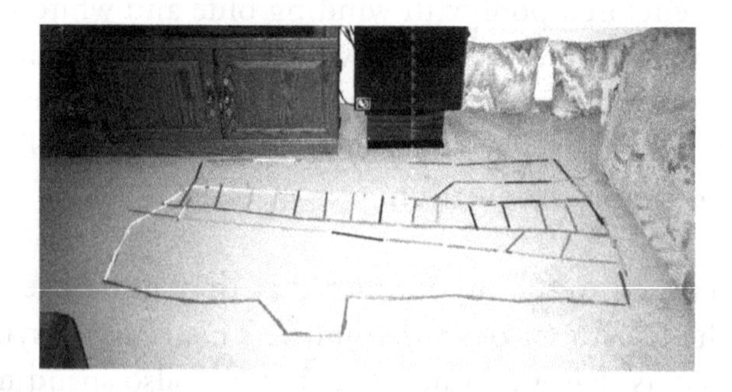

This is a picture of stuff that I would just line up, and no one could take something out of place or I could get upset.

When I had a paper that was completed (*or not completed*), that the teacher would collect at a certain time from all the students, it would be in my line of things and if they would go to take it, I would get upset. They therefore left it where it was and took it at a break when I would be distracted. Many people ask, "Why was this not reported?" I do not have a good answer for any of it, except for the fact it could cause the school to have to pay for evaluations, etc. On the other hand, maybe they just did not know back then, or did not want to deal with it.

I would also scuff my shoes when walking and tiptoe walk. I still do. This is another red flag of autism, as it provides input for balance. I am advised to have massages and joint compressions because of having to do this for the vestibular issues associated with autism.

At home, my parents would call my name and I would be staring at something or leaning my body against a sliding glass door. I was literally so absorbed in what I was viewing with my eyes, that my brain filtered out any presence of them calling my name from only a few feet behind me. The same sort of situations would happen at school and the teacher would call my name, sometimes loudly from right behind me, and I would not respond because I was absorbed in what I was working on. Sometimes the teachers would get frustrated and think I was doing it on purpose. However, they began to realize I was not doing it on purpose and would send me to have my hearing tested at the school (*which wasn't pleasant because of the high-pitched noises*). My tests would come out above normal. None of this was reported to my parents either.

I would also be caught staring at a fan or waterfall in the classroom (*also another very common obsession in Autism*) and then I would hear faintly, some people talking about me. However, these talks never lead to any action and I did not understand anything they were talking about, except that they were talking about me, by the pointing and use of my name. Numerous years later, after understanding my diagnosis and that there was something that I had neurologically, etc. I thought back to the conversations I heard the people talking about (*savant ability*). I could remember some of the words and *now* understand what they were saying. So, at times, I feel very saddened that I did not get the help sooner, because they would talk only amongst themselves and never to the professional or my parents.

During the school day, I would also be making some noises (*oral stimming*) and the teachers would keep on telling me to be quiet. They also realized after a short time that I could not control it. This, again, was never reported.

Similarly, during fire alarms, I would flip out and would hold my ears shut. Some teachers would have to help me move out of the classroom (*literally*) and out into the schoolyard. During high school, when things finally started to get a move on with a real good psychologist, I was taken out before the fire alarms would go off and be given time to transition. During the day, people used to notice that I would rub myself up against the walls a ton, and rub my cheeks and hands across it. This should have been reported to my parents or psychologist as I have now been taught it was out of the ordinary and is a form of tactile stimulation. I would also lean into bars a lot to get the sensation as well.

During the school days, I could not have my environment changed. Seating changes or alters would upset me and I would refuse to move. Because of this, they would let me have the same seat, as well as the other three people around me so that my closest environment would not be changed. I would not talk much and would get lost in the popcorn reading. The teachers had thought I was fooling around, but it was slow brain processing and trouble reading. On my all-star tests, I would almost fail most every one of them because I could never comprehend any of the books I read. (*When I passed by a few points, this was because I saw keywords I remembered 'seeing' and so I was lucky*) This concern was never reported to my parents either, as I guess the teachers decided I would catch up. Do you, who are reading this, think the teachers should have noticed all of these things as a whole and talked with the school psychologist?

I would take extra time on tests, and would be left inside the classroom alone, while others who had finished their tests would go to recess. I was so very upset and could not understand a word I was reading, or I was so slow, so I got scared and resorted to quickly copying some answers down or putting answers that did not make sense.

My handwriting was atrocious (*although my numbers perfect*) and the librarian would cause me to miss recesses to be punished by sitting in the library and working on my writing. It would not get better and the librarian and teachers would once again scold me. This is a very common fine motor skill deficit in both children and adults with Autism.

One thing I will add that will be explained in another chapter (*personal questions*) is how I would and still interrupt people. It is not on purpose.

I could go on, but I think you as the reader, get the overall idea of the problems that acquired throughout my years and that should have been evaluated long before they were. Please note, especially teachers who are reading this, signs need to be given thought and evaluated as soon as they appear, and accommodations need to be in place. I could have included some of this in my teachers and Autism chapter, but thought it was important to expand on in this section instead. There are many more instances of Autism warnings in my young life and throughout as I got older, but each chapter in this book will display what I have not given you here. (*Though, I must say, not one person could present everything in just one book – well, of course maybe the Bible is an exception!*)

2

AUTISM AND TACTILE DEFENSIVENESS

As some of you probably already know, Autism and tactile defensiveness is among one of the top concerns/struggles with the disorder. I know it was and is a definite struggle within my family with me. Clothing, socks, underwear, jackets and shoes are a constant struggle. That is not even mentioning touching something like a furry pear or oven mitt, or human physical contact. Then there is the food and swallowing problems, which most Autistic people have and dreaded haircuts. I am going to cover tactile defensiveness in my life, and because of it, what we have done to try to minimize the discomfort.

Shoes and socks:

When I was a child, my parents and their friends gave gifts of clothing, socks, little shoes and slippers, and it was a joy. However, it was only a joy for them. When my mother would try to slip socks on my feet, I'd sometimes cry. I would start to take them off because I could feel the seams and my toes could not feel the ground. It was literally too uncomfortable and unbearable. It was not uncommon for me to show up in sandals at school, which also occurred in me being disciplined for wearing them during P.E. At other times, I would show up with socks and shoes and remove them at school. I would flip them inside out as well, but it still made me physically feel sick. This is yet another sign that should have been evaluated. The socks caused tears and every time socks were even near me. The sight of them distressed me and I immediately connected to how they were 'bad' since they hurt me. Only since I have gotten older, through all the searching and anger in stores, were we able to find socks that were seam friendly and others that were seamless. An Autistic person will tolerate most seam friendly socks just fine. I am referencing (*check my website under tactile defensiveness*), to all of you who are in need of trying to get your child out the door, especially if you live in an area that snows, hence needing snow boots that feel just as terrible if worn without socks. The reference I gave above makes seam friendly socks for babies, children and adults. For completely seam free socks, I am now referencing (*check my website under tactile defensiveness*), although the other company who sells the seam friendly socks will work for most autistics. I also want to give another site, as it has many different colors, and gives all different sizes from little ones to adults and are seam free as well! It is, therawear.com, and you would go to SmartKnitKIDS seamless socks even though there are many sizes that could fit adults too, such as the crew size. You can contact them or call for more details as well! I am now able to wear socks without the severe discomfort and tears. For me, I replace the socks more often than neurotypical people would since they do get a different feeling after a while.

Undergarments like diapers, underwear and bras:

If your child or baby is constantly fussing, either their diaper or underwear may be causing them extreme discomfort, especially since Autistic people's skin is very sensitive. There are varieties of different diapers that you can try. I recommend practicing putting on and taking off the different diapers or underwear, and then recording the behavior pattern that occurs. This may be the answer to some of the fussing and tears. There are snug-to-fit, soft towel, silk cotton kind (*and because of the silk in them, are easier to transition into a pair of silk underwear which are much better tolerated*) and others. It can mean the difference between a happy baby or toddler without tactile defensiveness and a fussy baby or toddler with tactile defensiveness discomfort. A great resource for several types of diapers including the ones I have mentioned above, is (*again check my website under tactile defensiveness*) and have great service. For children and adult underwear, I would suggest (*check my website under tactile defensiveness*), and look at the microfiber or the tactel material underwear, as it is more like silk. Even when you look in the adult section, there are smaller sizes that should be able to fit a younger child as well. (*You do have a larger selection in the adult section*)

Bras:

Bras can also cause major problems and they really always have for me. I would recommend the (*check my website under tactile defensiveness*) that are like silk and have the tactel material.

Everyday garments:

It is important that the measure taken for these other clothing pieces be taken cautiously with everyday clothes like shirts and shorts, etc. I have had many screeching fits in front of people walking past in the mall and stores when I was younger, and even as I was getting older, as my parents did not understand that it was actually tactile defensiveness, but thought rather that I was just being stubborn. (*This is even when they knew I had Autism, as they obviously were not educated well enough. Sorry mom and dad*) I was actually in extreme discomfort and even the slightest brush up of a jacket cuff would flip me out. In addition, I would also cry because I knew what many people were thinking...that I was just being naughty, especially when I was high school age.

Food and swallowing:

I have always been in the lower sixth percentile in weight. The doctors were constantly saying that I was getting sick and constipated, along with low levels of nutrients and fiber, because I was not eating the right kinds of food. I was eating pudding, ice cream, applesauce, small noodles that were covered in slippery sauce, mashed potatoes, etc. I was not eating meat, fruits or vegetables at all, except the applesauce. I would sometimes have visits to the nurse at school because I would need a change of clothes due to this issue.

I went through very frustrating periods where I was going through tests to check my intestines, swallow reflux and esophagus. I would try to swallow a small cut up piece of carrot and I would either choke or gag unmercifully in everything that my body really needed. My chewing was more like gumming, so it would take me an hour and a half when it took my family twenty-five or thirty minutes to eat a meal. At school, in the cafeteria, I would have eaten about a fourth of my food and would be shaky throughout the rest of the day. I would hide food that I would duck under my desk to put in my mouth. Thank goodness there was a person who was diabetic in the class so it went unnoticed, but when it was, they really did not care. I guess they saw that I was different and saw me shaky.

I have always had a problem where I have food on my tongue, but it would slip down my throat when I was not expecting it to. This is because my mouth was not stimulated enough naturally to tell me when something was going to go down my throat, which lead to me choking and I would sometimes leak (*urinate*), which was both scary and embarrassing. As I have now grown older, I have learned through both great Occupational Therapists and our own experimenting, how I can get these much-needed nutrients. (*Autistic people actually need extra supplements as it is*) My tips are below:

1. Kid Co food grinder: This little thing is a miracle worker.
2. Z-vibe and tips: The Z-vibe and tips are great therapy equipment that helps the tongue learn to sense when food is going to slide down the throat. It is also great for introducing new food textures. It wakes up the mouth and the throat! In addition, use the bumpy textured tips for practicing chewing. The human jaw naturally constructed smaller over time so that

the brain in our species could be much bigger. However, an autistic person cannot compensate well enough with it.

3. Drink a cold or warm drink before meals to help wake up your throat. It also helps stop some of the tactile defensiveness in the throat that can cause a lot of gagging, etc.
4. Oral motor feeding therapy is using the Z-vibes and tips while introducing small amounts of food going from pure liquid to at least pastier or crunchier substances.
5. Because of these tips, I have been able to eat more than I was which helps to build up the nutrients to compensate with the other major issues associated with autism such as the immune system, other fluids and dental curbs.

You can get all of the supplies I have listed above in the tips from (*check my website under tactile defensiveness*).

I also want to add that I used to gum my food up and then I would stick my fingers in my mouth to pull it out or just mash it around. There were a couple of reasons for this. The first reason was that I didn't know where the food was in my mouth, so I had to physically feel where it was and make sure it was small enough to swallow. Secondly, I liked the feel and thirdly I wanted to see what it looked like after being chewed, as autistic people are very curious. This can unfortunately turn into pica, a disorder sometimes coinciding with autism, where the person will eat sometimes whatever looks like it would taste good or looks interesting. The body could even crave dirt and chemicals, which is why it is important to make sure that your child having their hands in their mouth at meal times doesn't turn into Pica but rather the other reasons I just stated. When I was younger at school, there was for some reason cigarettes on the ground, and I used to pick them up and eat them. I used to eat paper (*aside from ripping it all up*), amongst other things as well as taste substances. Thank goodness I did not have full-blown pica, but the level of pica I did have was not the greatest; no level of pica is anyhow.

Haircuts:

Autistic people's heads are very sensitive.

My mother has always said that she would never take me to a regular hair dressing place (*luckily she used to be a hairdresser*) because of the crying and screaming in pain. My head is very sensitive and the slightest pull feels terrible. The sound of the scissors start to get to me and the hair in my face and falling onto me causes me to need frequent breaks. Many times, hairdressers will use a paper towel or something similar to put between your neck and the cape. It is awful and I constantly try to grab at it, especially because it can feel like sandpaper at times. I am 22 now and still have major problems with haircuts. Sometimes, my mom will do part of my hair and then give me breaks, etc. both for my sake and for hers. I have given you some tips below that I have come up with (*as well as everything in this book*) that work not only for Autistic people, but also some that work for animals who try to bite you and get away when just brushing them.

1. Use a social story.
2. Buy and use a vibrating hairbrush for desensitization.
3. Have a mirror so the autistic person knows when their head is going to be touched and what is happening.
4. Use a visual timing device so they know when they are going to be finished, such as a light up and message timer.
5. Use the 1-2-3 or 5-4-3-2-1 technique. You do small combing sessions so they know that each time it ends rather quickly, and then give a short break. This I have found also works with dogs that are sensitive and try to bite when you brush them.
6. Let them be distracted with something they can hold and play with, such as a fidget. A great place for fidgets is (*check my website under adaptive sensory*).
7. Do not let the hair fall on them.

8. Let them play with plastic kiddy scissors so they are not so scared of the scissors when you go to use them.

9. With supervision and if safe enough, they can pretend or really cut a dolls hair.

10. Show them a video of someone getting their haircut.

11. Always be patient and sometimes only parts of the cut or trim should be completed. Then, you may need to finish the rest the next day.

12. Instead of just using paper towels between the cape and neck here, use something soft like a smooth knitted thing that stores make (*they go around the neck or something to go with an outfit*).

If you use these tips above, they should help. Many parents have told me that tip 5 has worked wonders for their children, and even dog owners!

For fingernail cutting, similar to hair cutting these are my suggestions below:

1. Take all the suggestions from the haircutting tips and incorporate them into the nail cutting process.

2. Maybe have another person in the house sit and you cut that persons nail in front of your son/daughter. Each time the other person in your house lets you cut one of their nails, they get something special that they like (*like a piece of food they like or something*) to try and encourage that as well. (*They need to look happy and rewarded*) In addition, do the same thing. It will take time but it can be very successful this way as well.

3. Sometimes, making his/her hands cold or warm before can help.

4. Also, massaging the hands with lotion (*it usually will become acceptable by many autistic people if not in the beginning*) as

well as massaging around where his/her nails are is good, so he/she gets used to that.

5. Don't use the word "cut" for cutting nails but rather "trim" or "make shorter...they're too long," etc. because a lot of autistic people take things literally. In addition, because of the literalness and slow processing, all the person may hear is, "I need to cut (*still processing*)"... then they get upset, as you could imagine.
6. For the sound of the clipping, he/she could have cotton in their ears while the clipping is going on. (*You may want to try cotton throughout the day so he/she just thinks that is normal and not exactly associated with the nail clipping, as it will be too much for them to process and they might take it negatively then*) The cotton can actually help throughout the day and he/she may end up wanting cotton in all day to help with vestibular problems associated with their autism and other sensory defensiveness issues such as louder sounds, etc.

Brushing teeth:

There are two sides to brushing teeth as well as tips.

Side 1: Some children will actually brush their teeth too hard and too long, because they need that input and it feels good to them.

Side 2: Some children will hate the feel and freak out.

What to do if the person is freaking out:

1. You may want to start with getting some z-vibes and tips and get tips that are flavored as well. This will get them used to something in their mouth that vibrates (*that can take some time but for many will become a wonderful soother*) and also have an extra good reaction by having a flavor they like all at the same time.

2. After using the z-vibes and tips, you can get the toothbrush tip to help transition (*that are also flavored, some of them*), and if they are not flavored you can put a good tasting safe paste on it to get them used to that as well.
3. After a while, a vibrating toothbrush can come along and they can have the same flavor paste put on that the child has become used to from the z-vibe and tips. Note that some of the oral vibrators turn directly into a vibrating toothbrush for much better transition.

Those are really the tips, and it can be a slow process, but is of course worth it in the end.

What to do if the person is brushing too hard/too much:

1. Get an oral vibrator with different tactile tips to help feed that extra needed input.
2. Do make sure you praise them though and let them know they are brushing very well. Also, show them all the different ways to brush your teeth so they can do it wonderfully and not just back and forth really hard. A role model example can be good and you can use this part of the tip for the section on what to do if the person is freaking out.

Activities to do for tactile rehabilitation:

I would like you to take notice of the notations that I make, but more specifically in tip # 4 for resistance purposes. I would also like to mention that you should not force too much, as you do not understand how it feels to the system – you can only go on created metaphors and should take them seriously. In the meantime, hopefully some of these recommendations can help! ☺

1. Goop – (CANNOT be eaten) Mix about a half a box of cornstarch while adding water gradually. You may also add

food coloring as this is optional. This goop will change from a solid when trying to pick it up and when squeezed, and then when rested in hand will turn to liquid. This can be kept in a bowl and added water to if desired to be kept for a week or so.

2. Edible Play dough! – (Yum!) Mix 1-cup of powdered milk and 1-cup peanut butter. Then, add honey for proper consistency. After that is complete, make shapes, create words or pictures for comprehension and language development, as well as of course fun tactile rehab, and then eat up. Hey, who said you couldn't play with your food?!

3. Pure baking powder – This can work wonders for tactile defensiveness issues. Just put it in a bucket where you can make a mess and dig in to feel and experience it on your hands and arms. You can also explain that this is used to help make foods and greater, show that what they are feeling and playing with is used to make sandwiches, etc. (*from the bread*)

4. Shaving cream – This can also work wonders and you can add food coloring as well! You can even explain to young boys the different uses, like cleaning and shaving. (**Giving explanation depends on the level – you can use a visual explanation - but it can help in the easing if they are resistant, meaning they could become less resistant if they learn and see that what they have in front of them is connected to something they enjoy, is important, or like!**)

5. Take some edible things like gelatin and put in a large baggy in the fridge. Then you can mix in shiny things (*like sprinkles*) so the child or person with Autism can feel it through the bag, which also helps to create visual awareness!

I feel it necessary to add here that tactile defensiveness also can be as simple as feeling too much air being cycled up into the air, and then into the person's personal bubble. Many times, the air reaches their face. At school assemblies (*when I had them and before I was excused from them on an IEP*) and when people walk by me in my house (*with socks on which cause more of an air cycle-up reaction*), I can become easily upset and hide my face. Some people or parents may think their child is being rude or just in a bad mood, because the face of the autistic person is actually looking uncomfortable and angry. However, it is the defensiveness of the sensory system dealing with the (*autistic person's*) sensitive skin.

I hope that you take all the information I have given you in some of the different areas of tactile defensiveness (*even more info. on the physical contact will be addressed in the autism therapies chapter*) and use them to your benefit to help ease these struggles and uncomforting tactile defense issues.

3

UNIQUE VIEWS ON AUTISM

To start this chapter, I would like to say that I have had many experiences in connection with my unique view on autism, some of which I have expanded upon, so I truly hope you can look at least for the moment past the clinical disorder and see the similarities to nature.

Some of my unique views on autism:

I feel that people on the autism spectrum are more like animals. Think of these things below:

1. Hamsters/ferrets move things in their cages into rows against the bars.
2. Horses get spooked if someone comes next to them unexpectedly, and have a different gait.
3. Cats get frightened when you touch them unexpectedly as well and do not always like to be touched. They also tend to love to squish themselves in small spaces (*deep pressure*) and like quiet, dark places (*a favor to many autistic people*) over being out free all the time.
4. Dogs are very sensitive to noises and smells, and can hear/smell things from far away.

5. Rabbits are very meticulous about what they will eat and what they will not eat.

6. Bulls are afraid/get aggressive when they see the color red: phobias and unexplained dislike of certain objects that are otherwise harmless.

7. Weather effects – Read the chapter on autism and weather effects, as well as view videos on my website.

8. Animals sometimes, if they are really upset about something will have an accident. Some people with autism, depending on the level, can have a similar reaction.

In all these ways, do you not think autism resembles the characteristics of animals? Maybe that is why autistic people are so close in their silent relationships with animals. There is a connection! I personally have a very close relationship with animals that I like to call my silent relationship. Is it not interesting that an autistic person can have great difficulty staring into another person's eyes (*I do too*), but more often than not be able to stare into an animal's eyes? (*That would be me as well*) I would also like to quickly add that eye contact in an animal's world is viewed communicatively as a threat or challenge. Therefore, many animals have a difficult time with eye contact as well, though with pending circumstances, can be tolerated and is wanted.

Nonverbal Communication:

I was walking by an autism room in a school and I stopped and watched for a few seconds until the classroom teacher, and aides, came to introduce themselves. One of their children, completely nonverbal (he had NEVER spoken a single word even through ABA or coaxing techniques, etc.) came up to the door. I stared at the little boy and we sort-of looked at each other. I sort of tipped my head and stared at him as if to say, "Hi," and he then stared back at me, tipped his head and verbalized "Hi." The teachers and aides were so amazed and told me that he had never spoken a word before. They asked me how I coaxed him into saying it. I told them, "I'm autistic; he's autistic... so that's how." People who say that autistic adults and children do not make relationships or connections are so wrong. Animals will communicate through various slight pitches in their tones and through unique body positions. Similarly, that is how we communicate as autistics when we are nonverbal. This is my own perception through research and interaction with nonverbal individuals. It is another similarity that our mind is destined to be able to utilize. The problem is that others who are not autistic cannot seem to get the hang of it, but we would not expect one to. It is just frustrating.

If you are interested in a wonderful view into the not-so asymptomatic DSM-IV criteria on autism, something that really makes you think, then I would recommend the book (*check my website under Unique Autism Views*). I strongly do believe in connection myself. Many times when I would be sitting there and staring off, I was experiencing this connection to what had been created in the beginning to be nothing but good. (*Autism and God Connection*) I also like to look at it as the God, autism and nature connection as well.

Autistic people are more like children, as in pure in heart and with childlike qualities. I, myself, tend to have that childlike behavior (*this is mentioned much more deeply in another chapter*) quality. Here is an example of a solution I came up with and wonder why it is not being utilized:

"If there is not enough money to go around, why can't the government just print more off to help people?" – Childlike thought

I do hope that all of you reading my book become aware of your self-thinking and note the similarities in your own children, and yourself (*that being said, assuming you are on the spectrum yourself*).

4

WHY HE/SHE DOES CERTAIN BEHAVIORS

Stimming is a popular word when describing the characteristics of autism. There are many stims that people know of, but they do not know neither the science behind them nor the inside view of why an autistic person does them (*unless you are autistic*), so I have provided them here below. It also explains why, on the contrary, you should NOT stop the stimming (*in certain times*).

1. Stim with their voice: Autistic people have used this mechanism since autism was first existent. The scientific background on voice stimming is given below.

   Stimming, most of the time with the Nnnnnn sound actually vibrates the person's inner ear drum, which helps promote balance and triggers specific brain waves to help relax and balance all bodily symptoms. Autistics have problems with balance and heightened awareness. Therefore, we use this method. Autistic people do not have to learn these stims. Their body is predispositioned to use these scientific methods.

2. Stimming with their nose: This is used as reinforcement similar to the hand flapping and finger flicking stim I will explain below.

3. Hand flapping and finger flicking: This happens much of the time so that the inner parts of the eyes that connect to nerves in the brain are able to be stimulated so that the autistic person can see things as they are supposed to be proportioned. That is why you may hear an autistic person say that things looked strange. Alternatively, also saying that they were so excited, it was as if their mind was racing and it was too much to take in. We are compensating for the overload. In order to rather view this as not just a part of a confusing diagnosis, I guess you can think of an adorable little chick flapping its wings. I just thought I'd add that, as to parents, it can be stressful and you can just think of my cute simile to make you laugh/ smile, if for even just a moment! (*I got the "if for even just a moment" from a commercial*☺)

4. Rocking: This stretches and pinpoints specific joint muscles in your back that 'manipulate' when the back moves forward, causing sensation throughout the body so the person can sense their body. In autism, feeling your body does not work correctly all of the time. However, rocking promotes manipulation of the joint muscles in order to get the 'feel' and 'grounding' in their bodies (*impulse*). This also works well to help autistic people relax and fall asleep when it is slow, especially when rocking while the autistic person is lying on their side on the rocking chair or person.

5. Peripheral vision stimming (*regular*): A person may move their eyes to the side. It is giving a balance between just seeing a small part of the object or world, which is less stimulation but gives a ton of input, and then gradually seeing the whole thing. It is like watching a train really close up. I guess you could also compare it to those pictures that trick your eyes, so that when you lean in you see the hidden picture, and when you lean out you just see a shiny and pretty picture.

6. Peripheral vision stimming two (*movie theatre*): Our human eyes see every square part of a movie scene on a huge theatre

drop screen; it is just that neurotypicals do not know it. There was even a study also linked similarly about the movie theatres using the unconscious/unvisual to the neurotypical eye process of the human brain to create condiment profits. They would use a flash production that neurotypicals couldn't at all see without a special screen divider, and with that have a picture of something like a Coke, a bag of popcorn, or even a specific box of candy. Before they knew it, people were starting to pour into that area and even leave the movie for a few minutes to go and get the item that was flashed on the movie screen. It caused their brain to crave these items on the screen. However, an autistic person sometimes sees these parts in the screen and at other times does not, but the brain does. Therefore, it is very overstimulating having to see the parts go together. What I am referring to is that it is like a graph, or sometimes just clear white lines in the film that are very light, so neurotypicals don't see that but Autistic people can (*depending again on the level and how the sensory system is balanced at the time*). So, a person's face can have a line going halfway through it. It can actually make a person on the spectrum dizzy in a way, at times, but the brain is still affected as well as the eyes. There is also the flickering from the screen and the 'blue light' that projects the movie that can sometimes be seen as well, so looking from the peripheral view allows those things to in a way, sort of disappear.

You should let the autistic person stim, as it is correcting the sensory in the visual field through the back of the head or brain. It is regulation that does need to take effect. But, do not let it go on too long, like more than a couple of minutes or so because then it can do the opposite effect of what it's supposed to do. However, after a few minutes of re-directing, if a meltdown or tantrum is occurring, it may be because the regulation was not completed. I want to add here that they are starting to now make Sensory Friendly Films in regular movie theatres for people who are autistic. Therefore, the family and autistic person can enjoy a movie; with the lights to their level, the sound not as loud, not so much action, etc. and still have it be the same movie! Now, I think that is progress and would like to thank the Autism Society of America for putting those ideas together and actually doing something about it. It is wonderful!

In this chapter, I am also adding the fact that autistic people are set off very easily. Many people who do not have a child with autism ask this question a lot, especially relatives or bypassers.

But another circumstance nearby can't set the autistic person off, can it?

YES, an autistic person is easily set off most of the time. One of my few special education teachers said, "Don't let those three sit next to each other. They will set each other off."

If another person is screaming or crying, the autistic person can be set off because it is their fight/flight reaction (*everyone has this capability so his or her body will be able to utilize itself properly in an emergency*). However, it is too high in the autistic person's system. It is normal all the time to see that when another infant starts to cry, your infant will too. The fight and flight system and body's self-concept is not yet balanced in their body. Similarly, an autistic person's body is still unbalanced in that area, even though they may be an adult. They believe that another person's phobia and worry must be logical, and therefore they need to "be scared" or cry too. At least that is what their system tells them. Therefore, their body starts pumping out the endorphins and other chemicals for the fight/flight reaction.

I am now going to mention some of my experiences with those stims that I listed and explained above.

One of my greatest friends ever, who is neurotypical, has said that she always heard me stimming with my voice. She said she heard that stim when we were at school and when we were near many people. I have also done it sometimes when I am on a chair, as sitting on chairs can cause autistics unbalance sometimes. I used to be made fun of for this stim, as well as all the other stims and effects autism creates.

My brother and others have mentioned and noticed the nose stimming as well and sometimes they think it is a tic rather than a stim, but it just depends as these things can go hand in hand with Tourette's. (*A part of autism many times*)

For the hand flapping or finger flicking, I had sometimes done it more when alone and more when I was really overstimulated. It helped me to see things properly as well as to walk better and balance better.

I used to get extremely upset from people at school and during a meltdown in my room; I realized that I rocked back and forth. (*It is not uncommon for drooling to happen during this, especially in the real younger children*) It was because during a meltdown (*as I will mention in the next chapter*), I could not feel a sense of where I was. Rocking would pressurize and give the input so I could feel "regular" again. I am set off very easily, in which service dogs help with that as well. (*That is mentioned in another chapter*)

I have had to leave the theatre for a bathroom break because of the separation in films, which if you were neurotypical, you would not see. I would also take breaks from the sounds, lights and movements in the movies as well. It is just peripherally overwhelming at times.

So, please, let your child regulate. It is like letting someone get insulin, so it is basically regulating him or her to "normal" for a while. To anyone who is insulin dependent, this is just an example to show that it is a regulation technique for the body and is important… it can change the physical part of the body for a while before needing more of the regulation. I, myself, have blood sugar problems really bad and have a meter. So, I sincerely hope that that statement would offend no one, as it is not intended to do that.

Why do stims change and over certain time periods?

There are two answers to that question, and yes, they are somewhat similar in nature.

1. Whatever affects the autistic person's brain is definitely going to affect which scientific stim the body is going to utilize.
2. It can be a repetition use to bring order to chaos. This is similar to the above reason, but this time the brain is utilizing the stim when it wants order to de-stress.

I hope you have found this chapter helpful in understanding further about the word stimming, which is often used to describe autism. I hope that you can now go and explain to others that there are reasons behind every stim your child has and that in a way, you can consider it well, rather smart if you want! The autistic person's body has a great instinct that can really help them and others in lots of situations.

In addition, if the person with autism is higher functioning, you can help teach them how to have some of their stims or tics blend in. For example, you have a stim and then sneeze to pretend it went along with it, or if humming, then go into a praying position. (*LOL, that humming and praying one was good☺; made me think of the special Christmas Monk show*) There are many unique ideas. You as a parent for the little ones and autistic people, who might not understand that concept can help that along if you want. It can go well and usually does not make the behavior worse thereafter.

Some neurotypicals can even use the first stim I mentioned (*the vocal nnnn stim*) silently in their brain or quietly vocally, which is explanatory, using it for vertigo or panic attacks. It can be very helpful and all you have to do is employ it softly. It certainly beats the full amounts of other time that would be of course otherwise unpleasant.

5

Meltdowns

Why do meltdowns occur?

You may have already heard that things overstimulate an autistic person's senses. Some of these things are as follows: The lights, the noises, the movement of people walking all around, etc. autistic people can hear everything; the pens clicking, someone's shoes scuffing on the floor, the voices, the lights actually hum, and even people's watches ticking. I told my doctor once that her watch sounded like tin when the second hand moved. She had to hold it up to her ear and could barely hear it. When the TV is muted, you may hear your autistic child saying, "Turn the TV off!" Believe it or not, when the TV is muted, the sound can still be heard. It is very quiet, but is still present. I was tested on this with some class friends at college, as I could not look at the people on the show so I could not know what they were saying from doing lip reading. You cannot hear all the words coming out as it is going through the filtering system in the speakers, but you can definitely hear some small individual words and the background noise of the different people's voices. In some televisions, it is more than in others due to age and where the sound comes from. They were surprised to say the least. I had some hearing tests done to see how good my hearing was. My hearing from the nerves in my brain came in much louder, and the hearing test people said that I could hear pitches that not they nor any other people could hear except me, other autistics and animals. (*Animals hearing have been tested and is analyzed by the shaking of their head and body language*) Sometimes, someone talking to me can sound like someone yelling at me. Your autistic child may keep telling you to stop yelling. You need to remember this explanation and just quiet it (*lol, don't mean to sound harsh there!*) even though to you it may seem too soft.

So how does it feel to be autistic when you are overstimulated? It feels like 20 cologne smells (*all people around you are wearing different things, etc. - Autistics can smell all of it and decipher them*), like hundreds of kids running around you and asking you questions in different languages, like you are sitting in a chair that is missing one leg and trying to balance, and lights are flickering. It is too much, hence why autistic people have meltdowns.

I had gone to a wedding, and afterwards I was crying for an hour until I fell asleep. It is just too much, but therapies can help along with some tinted glasses (*they are not extremely tinted, only a little*). They can look almost clear… hard to explain really. They are made for autistic people. In addition, if you have a fidget along, that should help as well. Service dogs are of course a major help (*I have a chapter on that*), and if there is a swing nearby they can also help a ton. In addition, fluorescent lights are awful. They feel like they are really burning your eyes, literally. They can also generate a lot of "humming." They are horrible lights. I can't keep my eyes open. I would like to bring up that in the cars at night, the traffic lights are brighter for the autistic person. Therefore, many times, I would look like a mummy in the back wrapped in a towel because it was too much for me to handle visually. In some people with autism, the light can even cause seizures. That is of course if they have a seizure disorder along with the autism.

Do they know what is going on during a meltdown? What happens and why?

With meltdowns, I try to explain what it feels like and why aggressiveness happens sometimes during them.

I try to describe a meltdown as this:

When an autistic person is upset, it is as if they go into a "spacey" type of mode where they are so overstimulated that their bodies cannot feel anything at that point. And, like others have said, "Snap out of it," to get the shock to stop, aggressiveness occurs in the autistic person as it is trying to fire feeling into their nerves. This is why sometimes they will slap themselves in the face, hit their own body (*even other people's*), etc. Biting may also occur during this, especially in the younger years. It can be from the anger, the inability (*if nonverbal*) to verbalize what is causing them to meltdown, as well as sensory causing them to meltdown. They could be needing input through the teeth, to the gums, and finally to the jaw. They are trying to feel something to get out of that sense.

When in a meltdown or angry, the person with autism may make a long string of sounds like lucklucklucklucк, etc. It creates a sense of relief or order.

Do not make the situation worse by talking even in the least. Sometimes you need to move away. Another thing is if you can get the person on a swing, and swing them, that can get them out of a meltdown. A service dog can be literally seconds. In addition, hand an oral vibrator to them and that will redirect the sensory system through the joints in the hand. It will redirect also through the head (*when in the mouth*) to help relax the brain tissues that are getting tense. Carry along other fidgets in your purse that could help to defuse a meltdown by assessing at home which ones have a greater "awe" effect, and which ones catch his/her attention the quickest (*and sometimes also the longest*). You can also have special muffed headphones (*special ones*) if the meltdowns are bad and frequent as well. They help because they create not only defusing of the several confusing sounds, but also apply soft, deep, and constant pressure to the sides of the head that tense up when the flow is out of control. I think you should also know that the person with autism could have a weighted vest or weighted belt. These help with grounding and feeling their bodies for overstimulation. It gives information to the muscles of the human body to relax the entire system. I do want to mention that the percentage of weight (*usually has scales on the links when you go to purchase them*) is different in the weighted vests and the belts. Some people love the vests and others can't stand them, as it can actually hurt their back or make them too sluggish, whereas the belt is there for balance and torso input which is wonderful because it is the middle of the balance system. A small piece of hook and loop fastener can be hooked to their pants pocket or something so they can rub that when they need to (*many times it's usually the roughest side*). Before I was diagnosed, I was given a piece of nylon hook and loop fastener by an aid for an autistic student in my class (*he was in a couple of my classes – I also didn't know what he had or what autism was yet*). She would also help me out, so I think they also knew then I had autism because she was being a personal aid to both him and me, but not others in the class. Therefore, when working with the autistic person be patient, quiet and distract. (*However, if the distracting is making it worse, just let the person stim if it is safe because they are getting out of it themselves*)

I will talk more about this and a few more examples of the ways and why's of what autistic people do, things they carry around, what they do with those things (*strings, etc.*) and what those things actually do, in the chapter about how it feels to be autistic. I will also mention a short story of when I got overstimulated, what happened and how I was evaluated to others of comparable age in that section as well. This is because I like to spread out my information, as I feel it makes it more interesting and more able to be read together as a group and discussed. I also have to admit that I like to spread things apart as well, because I would get overstimulated hearing/reading everything in the same topic area at once since I am autistic.

6

AUTISM AND LOVE

This chapter means a lot to me. All of the others have great value for me too, but this one I made especially to try and stop the sad rumors that they/we (*autistic people*) do not feel love, because that is not true. I will cover two areas in this chapter; the first being the love defined as caring about someone, and the second, romantic love.

To all of you out there wondering, because of hearing so many false statements about if Autistic people do feel love, the answer is quite simply this: YES!

I know what some parents are saying about the struggle they have with some of their children that seem to be arching or running away from their affection. I really wanted to lift your spirits in sharing these things below.

People with an Autism Spectrum Disorder really DO know what love is, I think sometimes even more than others. Autism can have at times such complete peace and beauty that it is like a little bit of heaven, hence autism and the God connection as well as my animal connection theories. Just the bark on a tree, running water, the howl of a wolf, the shimmering on a lake, and dew on the grass in the morning creates a connection that is based on complete love. Be happy and please be in complete peace in knowing that your child (*he/she*) knows love. Moreover, that he/she loves YOU so much. Someone can tell by looking into a dogs or cats eyes that they love their owner/parents. Sometimes they would rather just be next to or at least in the presence of them rather than in their arms, always licking them, or being petted, because the nerves under their skin are 5X more sensitive. This is also similar to autistic people. (*Another similarity between autistic people and animals*) The touch can feel terrible sometimes, and other times, just send the wrong signal to the brain. Any kiss or hug you get is great, but know that just because he/she (*the autistic child or adult*) may run or arch away physically from you, it does not mean that in their heart he/she does not or is not close to you. He/she is. There is no bond like a parents bond or one of a friendship. (*Please remember this, as this goes even for an autistic child, although there have been terrible rumors otherwise*)

I wrote a poem when I was younger, trying to tell the world that I could love, but when all these things about autism seemed to cover that message up. In the beginning, it may seem very sad, and you will see the similarities of autism (*the realities*) but you will grow to see the love of autism. Many parents have told me this is how their life is:

AUTISM – INSIDE

Sometimes as the time goes ticking by, I wait for someone to see me... Do not get me wrong people see me, just not the way I feel.

I do not really wish to talk to anyone; I just wish I could talk with someone.

I live my life just as any other would do, but my home seems to be unoccupied.

The little girl you see sitting in the corner, she lines her Popsicle sticks in rows.

And sometimes appears to ignore the fact that you have brought her something, but she's just not aware you are near her side.

At the picnic, she mistakenly slops the sauce all over the table. Why you wonder, how does a butterfly cause her to lose control?

The movies lead to bypassers staring, for the five-year-old covers her eyes because the light is so bright and holds her ears because it is too loud.

Her mother cries behind closed doors.

Her little girl isn't like the child she dreamed of having.

Her Daughter won't even bake with her. "Please don't cry Mother, the mittens just hurt my hands too much; I'll help you stir." That is what I want to say.

Your little girl screams and cries because socks and shoes traumatize her, so barefoot you take her out the door to ease her pain, and passengers stare back at you; "That parent has no control over her child," and I want to speak, "Please, please be kind to mother; you don't understand."

You're walking through a grocery store; once again, she has blocked the aisle by throwing her body to the ground because she wants to feel the cold grocery floor on her body.

She hides between those clothes racks and blankets in the mall, and then once again the clerks ask you to leave.

The parade causes too much movement, this time causing her to move her fingers in front of her eyes and make a screeching sound.

Why oh Lord, the mother asks, do you leave her Heart like this?

I dream that my daughter can fly high in the sky. I wish she could eat carrots and apples like her brother, but she gags.

She views the world in a different way, and I can't go into her eyes and see what the world looks like to her says the mother.

She catches a bubble and screams.

She even hugs the construction man she doesn't know because he is wearing a good-feeling coat and she goes to me but she pulls away; I wonder sometimes if she knows I'm her mother and does she love me?

Her memory so sensitive, she can repeat whole movies after seeing them only once and does as she runs around the room.

My dear child, how can I show you that I love you so very much?

"Dear Lord," the mother says, "I no longer ask why you let her be this way; instead, I now know that you picked her to be my little lamb. Thank you Jesus for giving her to me! Now I know she has a love like no other." (*And I now, the autistic person, can smile.*)

Now I am to the part about autism and romantic type of love. Some people on the spectrum, those usually diagnosed with Asperger's and who are higher functioning, can even marry. There are many people with high functioning autism and Asperger's who say either they want to love because it looks wonderful, or who say they cry because they feel it but they cannot live it. Some people, who are higher functioning autistic, will find another higher functioning autistic person (*who love each other*) and some may live together in a house with either relatives or the state checking on them every day to make sure they are doing well. I would like to add that there are devices for people on the spectrum like visiting angels, house aid systems, first alert, etc. so that the person with autism can have all those opportunities that others have. Service dogs can hit a button trained by psychiatric dog trainers for the person or individual with autism. Yes, an autistic person can feel romantic love. It is just different for every autistic person. (*Let me mention here, that the "sexual relationships" are usually none in an autistic person, that is depending on the level, for they don't find it necessary and it would be utterly painful for them*) Now, some people with autism who are AS or higher functioning and want a relationship are capable of taking care of a child together. With some help, they could adopt a child who is typical or a child with an autism spectrum disorder themselves. This is the other really good alternative. ☺

I mention in another chapter about autistic people living on their own, what each level can expect for living conditions, etc. but I find it appropriate to add a little something here in that era of subject.

I know that a lot of parents (*especially when they find out their child is autistic*) try to deal with the whole thing of "I wanted to have a wedding for my child, etc." but the thing is, some people are born to have that and other people are born to do something else; to teach the world in special ways. For those who are religious, I can say here that God must pick some very special parents to give a very special child to.

I now hope that from this chapter, you can feel relief in knowing that an autistic person can feel love, great love. Moreover, I do hope that you pass this message on and stop the terrible rumors.

7

DEALING WITH THE STARING

I am sure that most of you reading this know all about the staring, especially those places and people I mentioned in the poem earlier. In addition, I know that a lot of anger can arise and you may feel as though you just want to scream and yell at them or glare back nasty, and I am sure some of you have, but I am here to tell you another way. In addition, I'll add a few reasons why you might want to be extra careful the way you respond to these stares.

This is coming from me. I am diagnosed, high functioning autistic. Believe me, there have been many stares.

Therefore, here is my advice and a couple short stories of what happened in a few staring circumstances.

This boy had been staring at me a ton, and I was sure it was because I was acting different. At first, I was going to be mad, but then I just handed him an awareness card from (*please make sure to check my website under Emergency/Public Awareness cards*) and he said, "I know what this is..." I said, "Yeah? What do you know?" He then said, "I have a form of Autism, too. I have Asperger's Syndrome..." (*People with an autism spectrum disorder usually have issues with staring at others, not trying to be mean, but they find something very interesting about the person*) Both he and I came up with a thought that may make sense. "How would you know I'm staring at you? You must have been staring at me then to see I was staring at you." I guess this could be somewhat true! People on the autism spectrum like to think a lot. Now I am thinking that I am going to start being really careful of getting angry or upset. I usually don't get mad but see it as an opportunity, so that being said, I would say being careful would really be for the others who stick up for me, since you never know, the person might have a disability of some kind too.

In addition, I am thinking it is better to just have those awareness cards along rather than get mad, because then the other people who were told off for staring at your child, the next time they see someone with an ASD, they will have an automatic bad rapport with autism. Therefore, I have found personally, that those work the best.

I stare sometimes a little because I see a person and I wonder if they know their child is autistic. I will give you an example now of one of these instances. I met a woman and her son. Her son was throwing a tantrum in the bathroom where I was. I was kind of staring because I wondered if she knew that he resembled an autistic person in a meltdown. Psychologists tend to do that a lot too I think, lol, since it is their job. She got really mad and said to me, "Why are you staring at my son? Obviously, he's having a difficult time. You don't have to make it worse." I did not say anything. I pulled out a couple of fidgets, turned on the water, put it in my hands and leaned down to him. He started to calm down and his tantrum stopped. I looked back up at the mother and she was staring as if she was confused. She asked, "How did you know what to do?" I said, "Well it works for me and other people with autism…" Then she started getting teary and said,"Oh, my [gosh]. Well, I've had that in the back of my mind that he might be autistic, but I've been trying to convince myself that he isn't." Instead of then snipping at me, she thanked me and I gave her my email as well as gave her the same advice for if others stare at her son. They might be a psychologist, another parent with a child of an Autism Spectrum Disorder, high functioning autistic like me (*or have another disability*), or even just a person who is curious or does not know what is going on. Sometimes people stare too, although it is sad, because they want to make sure the child is safe… if you know what I mean. It is sad, but at least it is a reason. That is where those awareness cards really come in handy. It has actually helped some people who have then turned around another time and asked if they could help a mother who was in the middle of something, and whose child was having a meltdown. It really pays off to be kind back and hand an awareness card out, even if you feel like slapping them in the face.

I forgot to add, there are also buttons your child can wear along with t-shirts and zipper awareness tags. The site is (*check my website under public/emergency awareness*). In addition, the awareness cards are a great tool.

Moreover, from an autistic child's inside view (*my view*):

I have to wonder what is so wrong in that person's life that a disorder could cause them such anger. (*Of course, as I said in my original paragraph, there could be a vast majority of reasons, and good reasons*) I just smile at them and tell them what autism is. My dog helps alert people that I have a disability, so more often than not, now there is more staring at the dog! I am fortunate to have my dog, but still even before I had her, I felt the ways that I stated I did in the above paragraphs.

I personally used to get upset with my brothers if they got snippy at someone who was staring or pointing at me. Actually, it broke my heart, especially when I could not say anything or did not know what to say. I know that might sound strange, but it made me sad to see my brothers and people taking care of me getting angry with others who simply just did not understand or who needed the education. I felt like they were just as bad for doing that as the person who was pointing or staring. It also drew more negative attention on me, which was not pleasant.

I hope this chapter has helped you realize why you, yourself, really need to just let it go and give an awareness card because of the so many hidden reasons I listed above and talked about. This goes for even behind the anger or rude cases. So, please, for your child's sake and other autistic people (*as well as your own sanity of course*), use the suggestions above instead of resorting to anger yourself. You are not alone.

8

AWARENESS/SAFETY SUPPLIES

If we want to change the world and stop waiting for it to change, then we can start by informing people about autism and maybe changing some hearts!

In addition, we can teach the medical personnel and police officers what autism is too, so they can react correctly.

The awareness cards are a great thing to give out at holidays and Halloween, hooked to the candy!

The sites are listed below:

For awareness cards (*both medical and public*) I would recommend (*check my website under public/emergency awareness*). If you were getting a medical one, I would say that getting the magnet is a great way to go as it not only mentions what to expect autism to look like, but also gives a short synopsis and tips on what the person should and should not do with the autistic person, and how to handle things.

For things like buttons, mugs, lanyards, zipper awareness tags and taluses, etc. I would recommend (*check my website under the public and emergency awareness*).

Both these sites also have cards or awareness supplies for related and non-related disorders such as Fragile-X and Downs Syndrome.

Interestingly, both sites above are run by parents of autistic children. Go mom and dad!

For protection in neighborhoods, request your association or area to put up a sign, "Autistic child in area" or something similar. I have more protection tips for lots of places in your house, which I will add, as well as blind cords that are a common struggle with autism. I will try to create it in a picture format on the corresponding page on my site so that you can go and see the information as well! – That digital extra favor ☺ Maybe it will help you understand it more by seeing a layout. We will see how well I can do a layout though, LOL!

Therefore, I am now going to put in typing my information on safety and different parts of your house to help keep your autistic child/sibling safe from harm. Neurotypical small babies and children need safety as well, but when it comes to safety and autism, sometimes you need extra stronger things.

The bathroom:

As I will be talking soon about people with autism disorders having a 'broken internal thermometer', some autistic people could go to wash their hands and they could either freeze them too much or actually burn them, especially since they can get obsessed with the water and the feeling. You can buy a device that hooks to all the water handles (*also hooks right under the spout*) and lights up to tell you when the temperature is "ok," such as not too hot and not too cold. This is a very good tool so that your child or loved one with a "broken inside thermometer" does not run the risk as much of freezing, or even worse, burning their hands and/or body in water.

For children or loved ones with pica disorder, which can co-exist with autism, I encourage you to get a handle that attaches to the toilet seat.

I also recommend a seat that closes after using the restroom so that if pica is occurring, they cannot stick their hands in their feces, try and play with their urine, or worse. This also helps to eliminate of course the risk of drowning, as the child may become so consumed with the water (*always watch them longer during baths and assist*) that the brain can forget their head is underwater and they will start to breathe, resulting in the water going into the lungs. The space in the toilet and sink or tub, bucket, etc. is smaller which makes them feel secure and more apt to do that than a wide-open pool. I have safety things for the pool as well, some of which you may not have thought of.

Chemical areas:

I want to mention with the pica, which can lead to wanting to taste things that look good like chemicals or pills, that you do get extra secure "inside" hooks for your cupboards. There are also ones that are actual locks, where you have the key hooked to you. You do not want to be calling poison control or dealing with things falling.

All rooms:

As I mentioned in the above paragraph about things falling, autistic people are many times very adventurous, especially boys. They like to climb and reach even if it is unsafe. You can get attachments for your walls, furniture and figurines. It attaches securely from your desks, shelves, closet bars, etc. and hooks into your wall or ceiling for higher areas or lampshades as some kids will climb onto tables. This makes sure that when and if your autistic child climbs on these things or pulls at them, that it does not tumble over onto them and crush their bodies. This is so very important. Since they like to climb and their perception of height is not very well acclimated in their brains at young ages, some children will decide to go over your railing from a top floor or window shelf up high. It runs the obvious risk of falling and breaking things in their body. There are actual nets that are strong and look lighter in nature (*so not to arise curiosity*), so that when the child would go over the railing and fall off the top floor, the net would be there to catch them. It hooks into the ceiling and around the railing. You can use this for window panes as well.

Outside:

Similar to the net for safety in your home to prevent injury from falling, there are nets that are small enough and strong enough (*when the autistic person or child gets through the pool gate*) that will not end up in the water but rather when placed accordingly, will help keep safety. The net will instead have them sitting there just above the water.

Electricity:

I want to add also, that autistic people get very interested in things when they are told "no," and will want to test you. (*Some, quite many*) This is very scary for parents when they think of outlets. Please count every outlet in your home and get very strong and multiple step direction outlet inserts. This way, your child will give up. However, you want to catch them to reinforce. We do not want electrocution as that of course is awful.

The Kitchen:

For your autistic child and for other children, the stove can create a very big problem. There are plexiglass type things that go straight up and cup over the area where the burners would be as well as where the pots and pans handles would be. This helps to prevent the child from being able to just reach their hand in that area by standing in front of the stove. There are locks of course for the stove and for the actual buttons that turn it on.

I also want to mention that some children love the coolness of the fridge and some will actually fit in there. Then, the door can accidentally shut and that is a problem. There are special tools to help with the fridge and this issue that would allow the child to get food but not permit them to completely open it and then have the door accidentally shut on them.

There is a special tool as well for the sink grinder and a tool to stick and hook into it to prevent your child's hands from entering the area where the food gets grinded. This is very important, as it is a real safety factor as well as everything so far and to come.

Your child's room and baby's room:

I want to mention that people with autism that are behind and want/need to stay in a crib could fall out, so a cool, breathe-easy net to keep them securely in is both tent-like for the child but also very secure. It also helps keep infectious things out at night like small things that may be inhaled like spores that go through the air. The breathe-easy and keep-safe net blocks those things so they cannot reach your child who already has trouble fighting things off because of the immune system and imbalances due to the autism.

For the younger children, especially those who like to get oral input, you can put safe, plastic-type soft coverings around the sides of the crib so that they don't hurt their teeth or gums. This prevents bleeding and getting splinters/crib material in their body because of banging against it so hard or sucking on it. These are very important.

Your car:

Some of you may have not thought of this case scenario. Many children will try to get themselves unbelted from the car seat, as they are angry or do not like the feel of something. There is a lock on and cover so that they cannot get it out, or as easily. In addition, there are things for the doors; not just the general lock to lock all the doors, but special window coverings for those who are strong enough near a window to break it during a meltdown.

Rooms with blinds and cords:

Many children and people with autism are fascinated by twirling of cords that hang from window coverings or shades. This can cause serious complications like for instance strangulations, especially when they want to feel themselves wrapped up tightly. This is not even mentioning loops. You need to fix these blind cords and have them hooked down or raised all the way up and combined. There is a company that is dedicated to families called windowcoverings.org - Click on the left menu, "How To RetroFit." They will send you free retrofit kits as well. There are some small videos that give statistics so you can watch them on the main page of that site if you wish.

I add some more products and information on what to do if your child seems to always escape in a chapter later on.

One site that I would like to mention where you can get some helpful products for your autistic loved one is mypreciouskid.com

The other sites you will have to search for and should find pretty easily with the key words like 'stove cover, pool cover, railing cover', and some of the other keywords I gave you throughout that are not available in the site I gave above. I hope that this has enlightened some thoughts on things you had not thought about before and will help you and your beautiful child or sibling with autism!

9

Doctors, Dentists, EMTs, Hospitals, Firefighters and Autism

I am going to give you some background and then I will have many tips for these people listed above (our protectors) as well as parents of disabled children and people with a disability themselves.

As I have been writing this book, I have been sick for 27 months. You can look on my site to see the extensive details as a diary that I was typing (*while feeling good enough to write this book and type on the site*) to get an even better idea. I and my parents were asking many specialists to check for certain things that we looked up... therefore, we were the one having to be the doctor because some of these specialists were being lead to believe because of my record diagnoses, that everything going on was brought on by severe anxiety. Because of the autism and OCD. I had went undiagnosed with Valley fever for five months because of what I said above, as well as the hospitals deciding to give me some anti-anxiety medications and just sending me home even though I had symptoms such as coughing, a small rash, bad dizziness, trouble regulating temperature, nausea and fatigue. I was also very weak, had trouble walking and ended up in a wheelchair for traveling distances. Then, as time went on, I ended up almost completely collapsing after a church service. My eyes started shaking back and forth a lot more (*nystagmus*), I was throwing up more infection (*looked like grass- sorry but I want you to know the symptoms*), I started running high fevers after the smaller fevers and body temperature regulation problems, I started getting even more dizzy (which previously was thought to be caused by extreme panic, etc., due to my diagnoses), and was finally diagnosed late as having bad Valley fever. The Valley fever had got worse because the specialists and hospitals, were, as I had said above, looking at my symptoms as psychosomatic due to my diagnoses. This is even when I had some of the noticeable signs of physical sickness (*not mental*) during those previous undiagnosed months. Because of this late diagnosis, my asthma is about 85 times worse, *I mean resulting in many visits to the Urgent Care and Emergency Room, along with now having to take along a nebulizer when leaving home*. It has also resulted in my system having issues, so that when I am going out places, it is like I am almost likely to get anything that is around me. I had to have a spinal tap under sedation for the Valley fever to find out if it had actually went into my cerebral fluid as well. I came out screaming hours later and could not walk because of the procedure. It was not long after I had the Valley fever had went dormant that I pleaded to be tested for Strep, which I had to beg for two weeks for (*I would like to mention that everyone around me played a part in second guessing my true feelings because of my diagnoses, not just specialists*) to get tested. I finally was tested and it had turned worse as of course there was two weeks of it being left untreated. I ended up having to have special steroids (*legal ones not to worry!*) to keep my breathing/throat/heart under control. After I got through that, my system was compromised or sapped as you could just imagine and I was sure I had mononucleosis because I had all the signs of it. Different second opinion type specialists turned me away, so my mother took me to an Urgent Care when I could not get an appointment with my regular doctors. A doctor there decided it was worth a look after much convincing. It ended up being that I had to be extra careful of my spleen because it was swollen really bad since I indeed did have mononucleosis like I thought. Of course after that, just to my luck, we were hit by a car that was behind ours – not funny but just had to add that since the doc had said not to get into any accidents like sports or car accidents. The family doctors (*when I got an appt. with them after I saw others*) laughed when they were doing a routine checkup because I said, “Don’t press too hard or else my spleen will blow up!" Lol, yeah that was funny. They are doctors and knew not to blow it up ☺. To quickly lighten this up, one of my doctors that I think is cool looks like Richard Dreyfuss and my asthma doctor is actually on a commercial! Ok, back to the background: Not long after, I was turned away again because they thought I couldn't possibly have mono back again that soon. However, I was diagnosed as having it back again (*possibly because of my system being compromised from severe Valley fever and it being untreated for so long*). I was put on the codeine meds, which when I got home began having a very bad effect on me. Everything started spinning and I was choking on my own saliva. I ended up having the flu on top of that as well and was hospitalized once again. Even after my accurate conditions were shown, I was still questioned when I felt sick because of my diagnoses of autism and of course the big OCD. The specialists and other physicians wondered whether it was panic after Valley fever or whether it was really real feelings. The questions they had for us were always, "Where does the autism and OCD play into all of this, and then where does the real symptoms fit in?" After going through seeing about four ENT specialists aside from many neurologists and infectious disease docs, finally, the second to the last of them decided to put me on two medications. They had to do with the inner ear system to try and help with dizziness and lightheadedness, which again (*previously*), was pushed aside as panic attacks. The new meds have been helping quite a bit. I am finally getting some answers and results that prove that even though I do have these diagnoses, the feelings I am having are showing up in tests and they deserve treatment and examination. All the doctors at one place are talking with each other to try to come up with a plan, although I can only be seen by one for the major symptoms, because it would be too difficult for briefing and recording when only one is doing the majority of the work. I am doing vestibular therapies to help re-train my brain, which is a serious task, but obviously is supposed to help with repairing any damage that may have occurred. It will take quite a while and I may have to do it forever to maintain balance. I hope that it is not too late because of the constant sicknesses. New symptoms can arise from medicines that are so severe they seem more like an illness so I obviously start to get panicky; is it sickness or side effects from medicine? It is hard to decipher what is from new medications and then what is from a real illness. And again sometimes, there are both. However, they say I am not dying, but am just very sick and they are going to do more testing to see why I keep getting the same sicknesses back again and again (*some still unknown*: *trying to figure them out*). I have now acquired some extensive agoraphobia and PTSD as a result of all of this, which I am having to deal with. I'm also having worse tics, including breathing tics that are scary in Tourette's (*may be from the psychiatric meds…we are deciding maybe they are doing more harm than good*). Because I am not going a lot of places because of being sick and medication withdrawals, I am writing this book and am so glad to be writing it, because now all the pain I have gone through (*and am still going through many times*) and the important issues about autism that aren't really talked about, can be. I guess also it may be a proven fact that the doctors will investigate further that our (*autistic people's*) immune systems are on the fritz. I would also like to mention that I have found out I may have to have tubes put in from the damage and we will see how that turns out. Maybe it will be one of the pieces of the puzzle to some of the mysteries of my body. I sure hope that there will be more pieces figured out and soon!

Many people ask me, "How can you deal with all of that? Aren't you angry with the specialists and the others for continuously thinking that what you were feeling was in your head when it's come to be Valley fever and other illnesses?" My answer is this to those and others who have asked me similar questions, whose children or they have gone through similar situations because of having a mental health disability:

1. Am I angry that there was second-guessing and that I was not treated sooner, etc.? My answer to this is definitely, "Yes."
2. Do I sit around, stay angry with these individuals and wish I could just scream at them? My answer to this is, "No, because we are all human and it is confusing as it is just being alive."

3. What do I hope to gain from this? My answer to this is, "I hope to gain the attention of doctors, specialists, surgeons, etc. all around the world so that they can learn from what I've gone through, that just because someone has a mental diagnosis, it doesn't mean it is in their mind. You should do testing of the symptoms. Especially if there are physical symptoms, please try to look at them from a physical viewpoint and not a psychosomatic view. It is more than ok to treat the psychological point, but try not to just think of that area. Try to take yourself away from the panicky situation, such as detach from that and do what you were trained to do in medical school. Then, combine the treatments and testing together, even if those words OCD, etc. on the page are telling you otherwise.

These people spend years of their life dedicating it to helping people feel better, saving patients' lives, and performing tests to help someone else out while they run the risk of getting the sickness themselves.

I am grateful for my doctors and specialists, and although I do feel saddened sometimes, I very much appreciate the things that have been helpful that they have done and are doing. In addition, through the mistakes that were made, it showed the way to treat me even more correctly (*yeah, I wish this had never happened sometimes though, which is probably natural*).

It has also in my view, enabled me to give tips for doctors, EMTs, firefighters, etc., that if I had never went through this, I couldn't have provided for these specialists and safety personnel, families of autistic and or other disabled kids, and individuals themselves who have a disability.

The tips that I will give you below are from me (*as is everything else in the book*), so please use them and visit my website for both the videos and pictures to help explain them. I have added information throughout these tips that are not on the website, so be sure and read through them again even if you have seen some of them before, since I thought it was important some of these were included on the website for those who couldn't afford to buy a book. I also have some examples of how I helped someone in some of the scenarios, including scenarios that were similar to mine, as I have with every other chapter and will do throughout to make this worthwhile.

Higher functioning kids on the spectrum will probably freak out at a doctor's office or hospital just like other autistics who are lower functioning because of the tons of people, noises, unfamiliar things, etc. So many times doctors will say, "It's just anxiety from the autism and OCD… You can even tell by the way he/she is acting." No. If the child or adult with autism/OCD is complaining, then you need to listen to them. Autistic/OCD people can feel when something is wrong, and can actually feel it better than most because of the heightened sensitivity. Of course they can freak out in the hospital or doctor's office. This does not mean that what they are feeling is just in their head.

Hospital visits, doctor's visits and dentist visits, etc. can be a traumatic experience for any age person with autism. Many times, doctors do an automatic restraint system on the autistic person (*which makes the situation worse*) without trying other methods first, simply because they are "not aware" of autism and how to respond when faced with it.

A play medical kit may also be helpful. Alternatively, ones that look a little more similar to the real equipment may help prepare them better. You can find pictures and links to purchase these kits under the corresponding page on my site. You can also go to your local Walgreens store and into the medical aisles near the pharmacy. They have stethoscopes, etc. that doctor's in training use, or even people to help others they find in an emergency until the EMT's can get there. Therefore, that is another option if you do not want to wait for the more real looking kits you can order online. (*Not to mention shipping/handling sometimes – that is a whole other issue, in which I think people who are severely disabled or disabled under specific categories who cannot leave their house, should be able to get the necessities online, be able to punch in an access code and information, and get free shipping and handling set up by the government*)

When coming to an ER or other doctor's office, get into a quiet room as quickly as possible. Tell the people at the desk that your child is autistic and cannot handle all the people, etc. (*Depending on level, sometimes you'll just have to see as time will tell… however, if you think or suspect a problem, it's a good idea to just use this tip rather than regret having a bad outcome waiting to find out*)

I have been in a place where they had to get me a room because I was literally freaking out. For other places, I would go into a hallway that was empty and sort of darker to sit on the carpet. (*Which is an appealing thing to many people with autism – I'd like to mention here that this is another similarity between autistics and animals, as animals love carpet as well, only with them they don't rub it with their hands but instead lick it*) When I have had to wait longer, it can result in a meltdown that can be very frustrating for everyone.

I had to have an echocardiogram and was waiting an extra hour and ten minutes past the time, so I had a meltdown and panic attack. That was already causing problems. The lady was angry because she didn't know how to handle voice stimming while doing the echo, and the humming of the echo machine made me crazy, so the stimming got worse and I would sometimes need breaks especially for my asthma. She was extremely unprofessional and rude. She had stormed out of the room, pushed a garbage cart into the wall, threw her hands up into the air, cussed a lot and said, "I cannot work with a person who keeps doing things…" That put me into a worse meltdown. She got through about a little over half way and quit on me, which I didn't mind. The only thing that saved me from a nightmare meltdown was that on the way out, there was a cat near a bush. I stopped right in the middle of a meltdown and the cat came near, as I know that cat had that silent relationship and knew that I needed her. I was in the car and pretty soon, I started to have a mini meltdown, but the cat saved us from sheer screaming all the way home. On the way out, I had been so angry that I had been slapping myself in the face a time or two (*before I saw the cat*) and the woman just rolled her eyes. She never apologized.

The waiting, and especially having to wait around lights (*I was closing my eyes because they were burning and upset by the flickering*), the people, etc. was horrifying to my system. It is not a good thing for anyone, but really bad for the autistic person. This part is so very important. Most places have a hallway that the person should go to. Most people would want the person with autism to be away from others and calming down in our own way. I cannot stress enough how important this is.

Most places require paperwork, especially for new patients, so PLEASE request the paperwork be sent to your home instead. The waiting that occurs while doing paper work is a difficult and unnecessary time to wait and causes overstimulation. It also gives more time for anxiety and worrying about the appointment. The people usually recommend more time for paperwork, so then you would be sitting there doing nothing after the paperwork was filled out, quicker than they said it would take. My mom requests the paperwork to be sent to our home. Believe me, it's much better that way, even for the insurance information where they say if you are a new patient to show up about 15 minutes early before the appointment. (*And let me say some doctors do get behind a little so that would just be extra time added on*)

There is a great informational video for dentists. Check my website for the video. In addition, do send this video to your prospective dentist ahead of time, as well as the tips I will list below.

1. The dentist uses the weighted vests over my legs to calm me and to keep me from involuntarily moving.
2. Gustatory and tactile defensiveness reactions can make dentistry and autism very tough. Inserting tools or even looking in the mouth can cause gagging and sensation overload, as well as involuntary hacking of spit.

3. Make sure the suction thing is there if giving a shot for numbing or gel (*depending on the level*), because the tiniest drop on the throat causes choking and the involuntary hacking of spit. It is not on purpose... it is like a sneeze. This would also go for things like maybe shots or tests when the doctor might get mad for quick movements that occur at the time of testing or the shot. I had a shot once that had made me involuntarily move. The nurse got so angry and screamed at me (*my perspective but my dad said it was an angry raise of the voice*), which resulted in a meltdown. She did end up feeling sorry and apologizing. The sensory system does its own thing because the nerves are in charge. Sometimes, a calming gas through the nostrils can help to relieve these over hyperactive nerves that send signals to the brain to move. It is not the fault of the person, but rather the nerves and the brain. People (*doctors, nurses*) have to deal with all these things properly. Also, you may notice that your child's head moves really fast back and forth when they are taking a medication, and many times it is like they cannot breathe for quite a few seconds. This is involuntary and I used to get told many times, "Don't look ridiculous, you can stop this," but then of course it was realized later it was involuntary and very common with autism, again, hence the word "spectrum." I also want to mention that if the person with autism can handle the dental work awake, they are able to give a sign to the dentists if they need a break or more medication. The dentist and the parents/autistic person should set it up and even have it sketched out for all to help remember.

4. A small dosage of a sedative (*try the small doses first, as many autistic people have strange strong reactions*), should definitely help (*even for other types of doctors*). For more serious work, IV sedation may be an option you want to consider. However, the sedative pills can do an awesome job. Look further down for IV tips.

Print my entire medical page off from my website (*although unfortunately it doesn't have all of this extra important info. but at least has a general amount for them to take into consideration*) and give it to your doctors when going to a visit. Since you yourself are reading this extra information, you can help incorporate it into the process yourself while slowly teaching the doctors (*or scanning them off this entire chapter from the book and giving it to them with a note so they can read it in their spare time*). If it can be given to them before the appointment, that is preferable.

Sedation or general anesthesia may be recommended to make delivery of required tests or treatments possible in a safe and comfortable manner. This is for more difficult procedures and or treatments. The lower hemisphere can be up to 30 times more sensitive to autistic people, so procedures or screening in that area is TRAUMATIC. (*I am referring to gynecological type stuff*) This type of stuff has been attempted on some people with autism and actually resulted in severe dangerous shock, heart shock, the lungs stop breathing, and has even resulted in the worst-case scenario. These real terrible reactions have happened. I do not want to scare you if your child is willing to try it awake, but I just wanted this to be given out, as I think it is very important information and I would want to know, which I do. Some people with autism may be able to handle this, but many and most, cannot. I had an OBGYN that went on for 45 minutes saying that since I was verbal, I was going to have to have these tests done awake and not in a hospital, for it would cause her extra time that she didn't want to deal with. (*In the medical field, there are lots of things doctors will have to deal with and accommodate for. It is not a matter of choice many times, as you can tell by this and by the facts I stated above*). She kept on telling me this until I turned completely white in the face and almost passed out. She figured out fairly soon that if she did try any time, things were going to be thrown everywhere or my body would go into flight mode. Then, people could end up getting hurt from the autism reaction to emotional stress that can quickly turn into aggression without even meaning to. (*I would even get hurt*) In the end, we changed to a different one who understands and is very nice. They are even a religious based place and I can come back there several times just to get used to the place with my service dog. They are very kind, patient people, who take each person and work with them individually. They made sure to let me know that and also made sure that I knew we could talk about fun stuff too that I am interested in (*to get more familiar and just to be nice, etc.*). They will take all the time I need. Now isn't that how all doctors should be?!

For autistic people who are verbal, or who may be nonverbal but are definitely able to understand things, I highly recommend using soft words to describe techniques used by doctors. It is recommended as well for other children who are typical. Some of the phrases are listed below.

Blood Pressure Cuff: Give your arm a hug

Stretcher: Bed with wheels

Anesthesia or Sedations: Sleepy medicine

Stethoscope: A shiny circle that can hear

You should also have a visual pain scale during the appointments. The pain scale is better if you have faces (*want to add here to minimize anxiety, you can let the person add hair, etc. to them*) because numbers can mean nothing sometimes. We definitely are very visual people!

Also…

Autistic people can or may view the medical equipment as extremely terrifying, even just a stethoscope sometimes.

Autism/child friendly equipment on staff is also extremely beneficial and really does change the view to a person with autism, making the stay for everyone at the hospital or office a little less loud and upsetting. Check my site for pictures and links to purchase these autism/child friendly doctor equipment. It usually is not a matter of embarrassment for a person on the spectrum, because people on the spectrum have what others call "childlike qualities." These would be interesting and make everything a little less intimidating, as the autistic world is viewed much like an infant or young toddler many times. Parents or caregivers, you can purchase these items and carry them along with you for Urgent Cares, your own physicians, and even if there would be an EMT situation. They are very easy to slip on and really can make a huge difference. These slip-ons that I am talking about are like plastic characters (*mostly animals like dolphins*) and have a thing to slip onto the stethoscope or hook onto the tip of ear examination lights. A boy was being held down by five people, just so the doctor could listen to his chest. I told the mother about the technique above as well as letting the person look at, or have a similar kit at home during calm environments, etc. The boy does no longer have to be held down. She said that letting him experiment with them, watching other people use them, and having the child and autism friendly slip-ons for the equipment made the difference.

A person with autism disorder should have a psychologist/child-life specialist working throughout their stay at a hospital, even if the autistic person is an adult. Even adult autistic people (*again depending on level*) need the same sustenance as a child. Ask for one of them and a psychologist at the hospital.

Compensating with some testing:

When an eye is covered with a patch or spoon, the child or adult may react very negatively. Small scale bribing is acceptable to calm the person so that training or testing can be continued. Look at my website, under the medical page, to see a situation of a child with special needs having an eye exam. We learn what the child's near vision acuity is with single symbols.

One aspect of the role of a Child Life specialist is to teach young patients/older patients about medical procedures in order to make their hospital visits less stressful. They also let the children play with the equipment themselves. Look at my website for a small description and pictures. Your other doctors should do this, as well as explain slowly, and give the autistic person time to respond/understand. Many times my appointments are scheduled to be at least an hour longer than others because of the autism and its factors that play into this field.

All other people use the tips given below for these people as well.

EMTs and firefighters:

1. Autistic people will most likely be frightened of an EMT because of the following reasons: The vast majority of them appearing at once, visual-disillusion, meaning being frightened of the specific clothing such as the pants that firefighter EMTs wear since they are baggy and blow in the wind. It is a disillusion where too much sensory input is received through the autistic person's eyes.

2. When you are asking what is wrong with the person who has an autism spectrum disorder, be sure to use a quiet voice, sometimes talk very little, and when you do talk, use a robotic (*human but a little robotic*) or singsong tone of voice. (*rise and fall of the voice pitch*: *(sĭng'sông', -sŏng')*) This is very soothing to some autistic people, for they are heightened to repetition, etc. On my website, on the movie trailer page, you will hear this type of voice being used in the movie Silent Fall when the psychiatrist is trying to calm the autistic boy down. I will also add a clip of it soon by itself that is not graphic in nature (*as the movie is great representing autism but is about a murder*). In addition, sound like you are not worried but nonchalant.

3. Also, when asking what is wrong, point to the different body parts so the autistic person can respond better. It is visually helpful, because when autistic people get really upset and worried, even the verbal ones, they can literally almost forget words or forget what is hurting them. I have had this many

times and it really makes a difference. Sometimes they would touch my head or leg and it would help to re-fire what I need to help them help me.

4. When using things like thermometers, spirometers, etc., have another EMT doing it to their partner so the autistic person sees they are not going to hurt them. (*If you had the autism/ child friendly equipment that would be a plus*) They have done this for me in many places and finally I was calm and felt safe. Actually, we all ended up laughing together in the end.

5. Also, when ok to do so, talk to each other about everyday things to relieve tension. You can even ask the autistic person questions about something you figure out they like, to get their mind on something else so you can separate the anxiety from the main issue.

6. Always carry something, like a glitter wand or other cool distracter. This will help calm the person down.

7. If putting in an IV, try not to put it in the traditional middle of the arm since this place is EXTREMELY sensitive in many autistic people. If you can, use other places on the arm like the side of the wrist, which sometimes can be numb to a person with autism; if not when putting in, afterwards for most the time. In addition, look at my website, as there is a product that lights up the veins so you can find them easier. Then, you do not have to keep re-poking, etc. Also, this is a good time to bring up that you can gas the person with autism first (*a sleepy medicine mask – make it seem fun and have another person wearing one, but one that isn't going of course*) and then put in the IV so it is not an automatic trauma for the person with autism. This results in sometimes bad posttraumatic stress disorder since the system is highly different and sensitive. Also, this hopefully shows you that the person with autism stims with their arms a lot. So, moving the arm to stim will result in severe pain and screaming, panic, etc. That is another reason if you can put the IV in another area besides the traditional

middle of the arm, the better. Also, the fluids usually have to go in slowly at first when needed including meds. Be cautious that meds are very sensitive in the autistic person's body (*more so the relaxation medications but some others as well*). They usually have to give a "baby dose" because I have had baby doses that put me into not knowing who I was for three days, having to be carried everywhere, not being able to drink or eat, choking, and throwing up bile. There are other meds that were small and for anxiety that made me gasp for air, so they had to have a backup plan quickly. There needs to be signs (*big ones posted*) that say what NOT to give the person with autism, since the person is usually not able to accurately or quickly tell the nurses on their rounds to stop before they could put that medication into fluids, especially when they are groggy. I have had many nurses who almost put me in lots of danger because of the autism and strong reactions to meds (*again mostly anxiety meds*) that other people's bodies are less sensitive to. It can be dangerous, so several "HUGE" signs on the doors and above the bed, etc. are extremely important.

8. When the person gets to the hospital, immediately tell the doctors they need a psychologist on staff. Stay (*EMTs*) with the autistic person for a few minutes while they are being settled in. It will make them feel safer and get a sense of things being the same. Also, introduce the docs or nurses to them in a nonchalant way.

9. Either light or deep massage can help a lot.

10. The alarming sound of fire engines, ambulances, or police cars, are very loud and overwhelming to the autistic person's senses. Carry along cotton and/or earplugs so that you can place them in the autistic person's ears. The sound will then be a little less loud and overwhelming.

11. Be patient, also, if they are stimming like making noise. It is actually good for them, as it vibrates the inner ear creating balanced autonomic reflexes in the brain to coordinate balance

and self-regulation. Do not be alarmed. Go to my website in advance to see videos of these stims so you know what you could expect. I've had a couple people who didn't know what I was doing and looked freaked out, but luckily, one of the guys had a nephew with autism and explained quickly what it was. He was wonderful with me. All safety personnel should be trained in this area, and this is why I am writing this.

12. Firefighters, the person with autism may not want to come out from under the bed because of the smoke making things look different, and because of the alarm being loud. You should also use all these tips above and have a distracter in your pocket.

Talk to your local fire department and even EMTs about your autistic child and possible practice run through programs.

While I was in the hospital, a code red went off, the lights started strobe flashing and sounds were loud. I was screaming so loud and having a complete meltdown. None of the nurses who did not go to code red ever came in to help. This is unfortunate, but at least my parents had come back in, turned the light on, covered the strobe and sound as well as my eyes, etc.

The last time I was admitted to the hospital, the people were great with me. I had sent my website to the hospital along with bringing my magnet on autism and how to react, etc. with me. They had seen my site and looked at it more with me on the TV computerized screen with a keyboard. This is why I have sent my site to various departments that hopefully will take a look. I of course have lots more info and examples in this book and will eventually have on the cover who the reading material is geared for. So, that will hopefully attract the other important organizations that really need to hear my tips, etc., such as our many different types of protectors.

Also, sometimes a person with autism will get attached to a person working at the hospital. I did with one guy the last time I was there. He was given permission to spend extra time with me so my experience would be better. He quizzed me for knowledge, etc. because it was fun for me and he also watched "Bewitched" episodes with me.

Another guy saw that my anxiety was skyrocketing during asthma treatments. He used his charm and knowledge in autism to make my heart rate and anxiety almost go down to none, LOL, well the anxiety that is! He also visited me lots, came in and made me laugh, etc.

I have had other people who were absolutely terrible and did not want to even try to understand autism at all. However, I will not go into that because I want you to hear the benefits as well and the good experiences to give hope to those of you who ever have to go through anything in places like that. (*Again, as I mentioned in the beginning, use these tips even at dentists, regular checkups, etc.*)

There is a petition I started, along with a law in Congress that protects and causes hospitals, etc., to utilize sedation or FULL anesthesia for invasive procedures under Section 1557, including anything with tubes or needles etc. You are supposed to call authorities if the medical facility is refusing. They are supposed to have anesthesia teams on hand within minutes. *Visit my website!*

Please follow these tips. They are so important and can turn these experiences from disaster to okay.

10

POLICE AND AUTISM

This is very important, as I have had a police officer worry that I was either intoxicated or was going to steal, etc. He and his fellow officers followed me. That is where a service dog would be helpful, as they would alert the officers that the person is most likely not doing something bad. It would have been nice if I had had my service dog in this situation then.

I will address the full story about the situation and encounter with the two young police officers in Chapter 31, and I can guarantee that you don't want to miss that. Please keep a lookout for it and share with other police officers in training.

I know that in different communities, there have been quite a few problems with children with autism getting lost or running away. So, I recommend for all communities that these families with autistic family members get together with their town police force. This is so the faces become familiar with the police and help the people with autism to not be as frightened. This goes for firefighting as well as autism friendly drills the community can do together, and a guideline how, etc.

I also have heard on the news about the mistakes police officers are making, including tasing, because they did not realize the person was actually autistic. This is because behaviors of autism can be mistaken for other things that the police are supposed to be there for. The police should read these behaviors so they can clue into them. This way, they will be better able to decipher whether that person may very well be autistic and melting down, or autistic and lost.

POLICE: I am giving you tips to use when you have to work with an autistic person.

Read the Q's between the graphics on my website, on the corresponding page about police and autism. An autistic person may act out of control or strange, resembling someone who is high or intentionally doing harm in the community. In addition, some autistic children will get lost easily.

Police, also use some of the same techniques as listed in the chapter just before this one. Do not run quickly towards the person with autism as they will hyperventilate and could end up running out into traffic.

Also, try creating a community get-together to meet with the parents of children with autism spectrum disorders to get to know the child and recognize him/her, as well as getting the child or adult with autism used to seeing a police officer so they won't be frightened.

Check my website for a very good quote.

I would also like to add here, that quite a few people with autism who are higher functioning understand the general concept of stealing. However, there are things that the autistic person may be fascinated with that aren't for sale, and to them, would not be stealing but sharing joy. For example, I know it would be wrong to steal something on a shelf with a price tag, but there were several times that I would see taste tester spoons that are tiny. I loved the plastic and was obsessed with them. I was taking a whole bunch and this person started getting angry, thinking maybe I was stealing a bunch of price tag stuff as well because I was taking quite a few taste tester spoons (*this is also a more-than-not common fascination of people with Autism – the plastic part and definitely spoons*). They did not have a price tag on them so I did not think it was stealing or bad. There were people behind the counter and one of them wanted to call a security guard (*I found that out later*), but the other person had come up and just talked to me. She figured out that I really had no clue I was doing anything wrong whatsoever, as well as she realized I had a disability, and in that case it impaired my judgment. Or in other words, impaired my knowledge that it COULD be possibly considered stealing and something that would incur some worry. In the end, the women in the back thought it was too funny and unbelievable that I would love spoons so much. They ended up letting me have lots and every time I would come back to the grocery store they would let me have a couple for my collection (*I have a drawer full of different spoons*). I still have had problems because I am so fascinated with spoons (*as well as straws*) and I have to be reminded to stop taking them (*as well as hot sauce packages*). And, of course, in other places they figured out pretty quickly, actually laughed and said, "Oh well, at least it shows some innocence." It is like the slate in my brain from the first worry incidence is wiped away because my brain only has a "one track mind." It stops telling me the difference between price tag items and plastic stuff at a food place.

Police officers can be great with handling autism, as you will see mentioned in the chapter "Personal Questions," as well as the good experiences in school, which is in the chapter on teaching. They can be lifesavers and heroes to those with autism.

11

Personal Questions

These questions were given to me to answer by parents and teachers. I again have added questions, longer explanations and expanded answers. These are probably subject to be updated throughout the years.

1. What therapies do you think have helped you?

   Answer: Swinging, service dog, support group, ESAP class when in school, turning memorizing and homework into games, surgical brushing as well as joint compressions, rolling and bouncing balls, oral vibrators with z-vibe tips, a food grinder, reading program with pictures that can appear if you would like as an accessory, having a room when I was stressed in school with a swing, seamless socks and softer clothes and to finish, people allowing me to be myself and not trying to change me all the time.

2. Will you ever live alone?

   Answer: I do not know if I ever will be able to, but I figured out that it does not really matter because what really matters is whether I was a good person and did something useful in the world. No one has to be like everyone else, because what is the norm for some people is a different norm for others. For example: just like different countries have different customs, clothes and religions. I truly think that I probably won't. I touch on this in another chapter near autism and change. I also want to mention that it might be possible with a service dog that is trained from a very good organization that deals with independence assistance scenarios.

3. In school were you in mainstream classes or were you just put in Special Education classes?

   Answer: Both

4. Will you ever drive?

   Answer: Maybe just small distances with only little traffic. But personally, I don't want to and I don't think it would be safe for me due to my phobias. In the future new options may become available!

5. Do you believe in all the different categorized names they have within the autism spectrum?

   Answer: I do not think it much matters. Either you are not on the spectrum or you are. However, with the label autism, you get way more services that are crucial in all levels.

6. Did improving in some areas come natural to you or did you just outgrow some things?

   Answer: I have heard that people on the spectrum develop at a slower speed in some areas. Therefore, what a typical twelve-year-old can do, a person on the spectrum will do at a later age such as 20 or even older. This is especially true within the areas of emotional/social and other needs. I have grown in some areas and have gotten worse in others. Look through my book and you will see that I explain in here about the different stages of development.

7. Did you find that school would upset you and you did not react until much later in the day?

Answer: At school I would be a disaster, and then at home was better because I was so relieved to be home. My family never knew the stuff that was happening to me until much later. As I got older, I would have meltdowns at school that were worse, like kicking garbage cans, storming out of class (*either from mean kids or a couple teachers I had that were horrific, even a special education teacher*), etc. I would calm down with the psychologist or officer in their area or swing. Then, I would also freak out at home as well and breakdown there too as I got older. My parents didn't always see this happen as I hid in my room. It is so much stimuli and the one thing that helped me was this ESAP class. We were all the same, or at least similar, and became a supportive family. We all would meet 2nd hour and I had a few other special education classes in that same room. It was a space for me to go and calm down too during the day. I wish every school had that specific class, as the psychologist was in there the whole hour at times. It was so wonderful and that definitely helped. Sometimes, I would freak out even in the younger years when I would get home but it became more common in the older years, like junior high school and high school. It is not uncommon to have a child with an ASD go through school and blow up at the end. It is a result of all the stimuli and bullying building up… like a hose that has a cap on it and finally explodes off. Similarly, it is possible to have a child blowing up in school and calm at home due to relief.

8. Did you find that the related OT helped you cope better in school?

   Answer: Oh yes, the psychologist would do O.T. on me and the aid would do deep pressure, etc. In addition, the great nice police officer who had a psychology degree would give input all throughout the hallway swinging my arms, doing deep hugging, and walking with me to all my classes. He would pop in also to see how I was doing in the classes! In addition, the psychologist would pop into my classes too sometimes.

However, the police officer popped into my classes everyday just because I loved him so much! The other classmates would laugh because he would be funny. I do not think they thought he was there to help me out for psychological reasons, but just thought he liked to choose me while entertaining the class for a break.

9. I am doing surgical brushing with my child during the times our OT has recommended doing it according to how her "engine is running." However, when I stop sometimes, especially at night, she starts crying and freaking out wanting me to keep doing it. The craving could last for almost the whole night. Do you have any ideas as to how to help with this situation?

   Answer: Yes, I do have a suggestion that will hopefully help you. I have had that same problem and it's like I will pull back on the person who was doing the surgical brushing or joint compressions, but especially the brushing (*as it is very, very tactile!*). I would pull them back and get very emotional and meltdown-ish as well. What we found out worked for me was that I would have my fingers rubbed by it, and eventually, I would just end up rubbing the brush with my fingertips until I fell asleep. It is a security and a tactile input controller to help so that there is not input and then an automatic stop in the system (*like getting food and then going into a blood sugar drop*). I hope that will help you. Just try rubbing it against her fingertips and quietly with the lights low, begin to let her do it herself. It should help her to fall asleep with it next to her and she can get that little extra input so it is not a drop in her system.

10. Did you find imaginative play to be an issue growing up? And if so, was it something you eventually learned?

    Answer: Oh, I had major problems with imaginative play. I used to take all the pretend food and the pretend stove and kitchen, take the foods and put them together (*make a hamburger or hotdog, etc.*), and then line them up in the cupboards. Then, I would take them out and re-do that. That could go on for an hour. I would do the same thing with the dollhouse or Barbies and their house. I would clean everything... position it properly and that would be play. I did not understand why you would speak for a doll or pretend to feed it, because they are plastic and not real. I still to this day, with my two little cousins (*who asked me to pretend play*), did not know how and I did not understand it at all. I think I looked at them as if I was confused, so they just played together and I just sat there uncomfortably watching. This type of play is so difficult, and sometimes, in some or many autistic people, does not ever become accomplished. However, in some autistic people, the imaginative play kicks in late like literally about the age of 25 or 30. Then these autistic people at those ages are buying baby dolls, dollhouses, and playing. It is like a deprivation and then they want to see what it was like for other children who had that experience. There are real disorders out there, a couple, that are similar to what I have stated above. They have to do with psychosomatic issues or very important formative years that were not formatted or fulfilled properly. (*This can be thought of as someone who is a small child and who was never spoken to so their speech is not there, or a child who was always carried did not learn to walk, etc. – of course this is a bit different but this I thought would just help*)

11. How do I get my child to be more engaged? He seems to just want to ask me for things all day and talk about what he wants for requesting. How do I get him to want to talk to me? How do I get him to stop pacing and scripting and recognize his family and me?

    Answer: This is a bit tough, because with autism, it is a much-scripted thing. All he really knows is when he needs something and that is what it is like for him. However, here are a few tips. Find what he likes and you yourself engage with him, which should in turn, after some time, have him engaging more with you. In addition, sign language, child sign, is a great engager and promotes speech and interaction! There is a site where you click on some of the words and then there is a video of that sign. It is (*check my website*). I want to add here that it is a good thing though that he is asking, and you should praise him for using his words to express what he wants. That is a great milestone and you do not want to break it by trying too hard at the other side of engagement. So, go slow and it should work out fairly well and fairly soon.

12. When you are upset, do you know what is going on around you?

    Answer: I sort of know what I am doing, but being on the spectrum, it all kind of happens and it is hard to explain. I do not always know what is going on when I am really getting upset. It is kind of like when someone says, "It is like in the movies when they make things happen in slow motion or when they speed things up or rewind to make a point." It can be very confusing, therefore making the brain see but not filter what is going on at the time until much later. This is when you can talk over what happened at a calm time. You can then mention an alternative that they could use during a similar situation and show signs to help them notice if a situation is going to be similar to what they just went through.

13. Did you only have problems with haircuts as a child or now as well?

    Answer: The haircuts I really have a terrible time with, even today. My scalp is sensitive, which is not uncommon in people with autism. In addition, the hair falling onto me and the sound of those scissors is terrible! My mother dreads having to do a haircut as well as me, so we love the time in between in which we don't trim. I want to mention here that conditioners help a ton we have found out, because it coats the scalp and helps for lessening the amount of tangles. I want to mention that you can wait less time in between so that you can do quick little trims that do not take very long. That can help to minimize the dread of the haircut, the time it takes, and result in a better outcome.

14. My child keeps getting upset, keeps talking about the lights upsetting him, and many times seems to be concerned. What could be the reason for this?

    Answer: Many people with autism have eyes that are so sensitive to light rays, as the person with autism can see rays coming out of lights that typical people cannot see unless perhaps they are looking through a prism or 3-D special glasses. It is not that extreme of course but we can see a couple rays from each light, so it's very annoying, especially in some specific lights more than others. Also, lights because they radiate heat, create a sort of stench or smell that again typical people wouldn't notice but an autistic person (*another similarity between animals and autism*) can smell very well. It can smell like a burn type of smell or like a fire type of smell. I have many times said or thought to myself, "It smells like fire in here." I have told that to my dad before and he has assured me there is no fire starting anywhere; it is just me again. My animals and even my cat, especially since she can get up on the shelves, sits and smells the lamp under the shade, which helps the explanation be clearer here I hope. As something funny, I want to add that her mouth opens, stays open for a while, and it looks weird!

15. My child is not too fond of sitting with the family at the table during meal times. Do you have any reasons as to why?

    Answer: There are a few reasons. One thing is that some autistic people can get very unbalanced in a chair due to the vestibular system, which becomes very uncomfortable. Therefore, you will see them sitting on couches or other areas other than chairs. Also, the sounds of other people eating, like the crunching or sucking of things, seeing spills happen that aren't cleaned up or seeing foods that bother them (*OCD or phobia related*), etc. can cause extreme problems there. Therefore, the autistic person becomes overstimulated and wants out of the situation. They just want to eat alone, as they get enough sounds in their brains from their own eating. I want to mention here that some people with autism (*many*) have a problem with specific food sounds. One kid who was at school, his mother informed me that he would get agitated when others would be eating a pretzel around him because of the sound. The aid had been eating pretzels so he kept becoming agitated (*she didn't know at the time but found out later*). The aid actually used it as a punishment, which is always a bad idea, because you are overwhelming the system that is already overwhelmed. So, I am explaining it may be from specific eating sounds and specific food sounds that he does not want to be there during meal times. *I want to add here that taste aversions are sometimes used to stop some unwanted behavior, like hot sauce for example, which is popular within that aversion practice. It can be replaced with other things I think, but sometimes it can do the child good because it sends the strong taste of the sauce (*the spice*) through the nerves to wake up or correct the brain. Basically, it is showing the brain what is going on and the autistic person just responds.

16. My child scripts a lot. How can I encourage him to stop doing this?

    Answer: This is very difficult and I really wish I had a helpful tip for you, but on the other hand, I do have some information that may help you look at scripting in a good way and ultimately may surprise you. I, myself, since I am on the spectrum and on my level, have issues communicating and understanding topics other people talk about so I script. I hear many responses in conversations (*because of just hearing it like others but remembering it because of the savantism and the conversation around it*) as well as hear things from movies and TV. Then, when I am in a situation where I do not understand really anything people are talking about, I will hear some keywords that I heard in a previous conversation or show and remember a line or phrase that was used. I can then interact with the group and in a way "act" as if I understand what they are talking about even if it is only because I am scripting. Sometimes (*quite a lot*) I will get very lucky because my remembered script will go hand in hand with what the others are talking about. However, at times, my remembered phrase I will script out to interact with the group really doesn't go along with what they are saying. Then, I have to explain what I was doing which makes the group ultimately intrigued or it can go the opposite way. So, scripting can be very beneficial. I'm sorry I didn't have the other side of help to offer to you as I don't know how to stop myself, but I hope that what I said enlightened for you another way to look at scripting and the benefits it can have!

17. I will ask my daughter a question and her response most of the time is, "I don't know." Why is this and can I possibly do something about it?

    Answer: Oh, this is very common. This is still a mystery somewhat, as the autistic person's brain has the answer to your question but automatically comes back with the I don't know answer. This is similar to when an autistic person who is lower functioning says, "Yes," when they mean no and, "No," when they mean yes. Also, sometimes it is because the brain is tired and they do not want to think anymore or be questioned. There are things you can do. You can have a paper they can fill out later with your written questions or you can give them short questions. For the paper, sometimes you may need to have a "Yes and No" so they circle which one answers the question. This is easier for them than having to write it out. Also, for example, you could verbalize, "Did you eat lunch with a teacher today?" Or rather, "I bet you had fun eating with ____ today." It still can be the I don't know answer, but this will usually come along with the help of those tips I gave above and with the chronological age in autism. Also, you can act excited and as if you already know what their answer will be. I hope that helped some. I also want to mention here that people with autism will be doing something like getting food ready and will ask a question. However, they already know the answer and are already getting what they need. Then, they'll ask again as they are performing what they know but are asking as if they didn't understand. For example, they might ask, "Where are the spaghetti sauce packets?", but while they are asking the question, they are already on their way to get them or have them in their hand. This is a mystery indeed and there are no great tips for this yet or understanding. I thought you might find that interesting as well, as it is similar to your question, and you may find that happens too!

18. I know that people with autism have very sensitive skin and ears, eyes, etc. Can you help explain it more to me besides the same diagnostic explanations I keep getting from the doctors, like a metaphor or example?

    Answer: I do have a metaphor to help you understand this hopefully much better than ever before. I am sure you have heard of Helen Keller, the child who was born blind and deaf. When a person is deaf and blind, their sense of touch is heightened and in a way, their ability to feel things that a non-deaf and non-blind person would be able to feel raises higher on a tactile scale. For instance, a person who is not blind or deaf can feel threads and textures on a car seat or other furniture. However, if you have this impairment (*blindness*), the nerves react differently in your hands so that it can compensate for the blindness. Basically, your sensors in your fingers are becoming your sight for everything around you. Then they can pick out different threads and the patterns in them that a person who is not blind couldn't do nearly as spectacularly. When a person is deaf, they become aware of feeling how they are walking and feel every bounce of when they walk, as the feet and muscles become the sensors of hearing. I want to mention here that animals who do not see, like in the ocean, rely on these special heightened sensors that compensate and become their sight and hearing locaters. In autism, it is as if the system is blind, deaf, etc. So, as you can imagine, each sensor system is fighting to take over and become the alpha sensor for the other systems. It doesn't realize because of the vast brain differences in autism that the person is not deaf, blind, etc. I hope this is making some sense. And, in another sense, hope this makes you see this connection again between many autistic people and animals.

19. My child with autism freaks out when I ask him/her to clean up their room and he/she seems to be precise in some things, but yet so 'in many other areas. I always thought autistic people liked things lined up, so why is this so?

    Answer: Your child with autism is probably freaking out because when the person with autism is looking at a cluttered area and then hears the words, "clean it all up," believe it or not, they have no clue where to start cleaning or where things are supposed to go. Their brain does understand this stuff, but they get so overwhelmed visually that they can breakdown. In addition, with regards to organization, a person with autism can definitely like things in lines but also in patterns. What they/we as autistic people can see (*photographically*) is very different than what you are seeing as a neurotypical person. I am going to help you here as to how to defuse the anxiety and confusion from the cluttered room and the cluttered mind of your autistic child. First of all, get down to their level (*like leaning down to reach their height*). Say to him/her, slowly, "I know this looks like too much to clean and you probably don't know how to start, right?" Give them some time to answer. Sometimes you can give them the answers that they want to say but are not verbalizing like, "Yes, it is… I know that you think that and I'll help you." This gives the person with Autism some relief to hear "I'll help you," so they don't feel alone. Then, help him/her clean some of the stuff in their room and give rewards as you are doing it. You can even have him/her watch you put a few things away and then ask him to hand you some things he/she really likes (*with a smile*). THEN, you can tell him where they should go and simply place them there. Be sure and give praise, but not overstimulating. I know it seems like he/she needs to do it by themselves, but if it starts that way in autism, it can result in big meltdowns and further resistance. The brain is overstimulated, so you need to help de-stimulate it through the techniques I have just given you. I also want to say that when the room is looking a little cleaner at least (*sometimes take breaks and let him/her know it's ok to take breaks*), then while he/she is or isn't looking, you can take a few things and put them on the floor. Only put a couple things and they can be more of the child's favorite things in the room. Then ask him/her where they think their favorite things should go. LET him/her know that it's ok for them to have time to think. If they get it wrong, don't act upset but just say, "Well, that was a really good guess, but here is where I would put it." Then guide him/her with their favorite item and have them place it where it needs to be. Try and make this experience as educational, de-stimulating and as fun as you can by creating pure and positive things. These things would be perhaps remembering games, rewards, praise and even favorite toys for learning!

20. My child with autism hates getting water splashed on his face, yet strangely loves to swim a lot. Any reasons as to why this is?

    Answer: A lot of autistic people do not like spontaneous splashes because they don't know when they are coming. It can also be a transition issue. Many parents say, "My autistic child is like a fish!" This is because water is very therapeutic and can help ease tension from muscles and tones, especially in young boys with autism. In addition, it is not uncommon for the liking of swimming rather than splashing because the body is submerged within the water. I, myself, hate getting splashed in the face or anywhere else as I start to get overstimulated, sometimes stim and then I won't be able to have fun at all. When I have been at church Luau's, the leaders always had everyone know when I was getting in the pool. They told them to tone down on the splashing around me and do more calm swimming so the splashes weren't like "BAM!" and then my brain freaks out. That is why he/she most likely does not like splashing but yet loves swimming at the same time. Don't forget it can be a transition issue as well. I give you a good explanation of this later on in my book about your child and transitioning in and out of bathtubs.

21. Do you agree with doctors giving a life-long prognosis so early in life like around age 2 or so?

    Answer: Sometimes if the child is truly profound, then the doctor is sending out a "warning" so that you can take classes to help prepare a good future for the child. However, with the therapies, medications and technology, it's possible for a very autistic person who seems profound to become fairly functional. Also, you do not want to just give your child the minimum acknowledgment of what they can accomplish when he or she could deserve much higher credit in the years ahead. (*Meaning you just accept that at 2 years old that is all they are going to amount to when they could progress*). So, I agree in the form of "warning," but then again do not agree. I hope that makes some sense to you all.

22. I feel so guilty about my child having autism and cannot seem to come to terms with it even though people say it is not my fault. Do you have any advice that might be perhaps something I have not heard before? I don't know if you can, but maybe even reassurance?

    Answer: I think I may have something that people have not just said to you repeatedly, like, "It's not your fault, don't blame yourself." It is not your fault first off, which you have obviously heard and probably been told about through various theories that cause autism. However, I have a different way of looking at the word "guilty" that you or others trying to help you have not seen. Have you heard of that phrase, innocent until proven guilty? Guilty would then mean that you intentionally when you were going to give birth said, "I want my child to be autistic; I want my child with autism to have vestibular issues with walking; I want my child to have autism so he/she can be sensitive to everything around them in the world..." I hope you are getting my point. You cannot possibly be guilty of anything or really feel guilty, because "guilty" means intentional or pre-meditated. I hope that gives you relief. So, no, you are not guilty and you cannot feel guilty (*I know easier said than done*). See, when you look at it that way, guilty truly meaning intentional or pre-meditated, then you are not allowed to feel that way. You will have to come up with a different word like feeling sad and from that point forward, continue with the therapies, love your child for who they are and flourish with the beautiful gift God has given you.

23. My child does not seem to know at all how to entertain herself and just keeps bumping into all of us wanting constant attention. Do you have any ideas?

    Answer: I do have some ideas and possible reasons for this. You mentioned that she bumps into you guys constantly, so I think her body may be needing deep pressure for the autism sensory needs and joints/muscles. You can try pressure or weighted vests/belts as well as joint compressions coached to you through an O.T. The link to these important things are on my website under the section adaptive sensory, so be sure and check them out. In addition, many times a person with autism has something they will find as an obsession. Some people think at times it is not good, but many times it can be good for entertainment and even jobs. For instance, you can try a couple days to find that obsession such as specific series of shows, animals (*she could brush, feed, pet, play and take care of an animal*), trains, something in science, water, sparkles, gems, puzzles, talking books, etc. Then, she can start to do them in the beginning with you, but not for too long. Before you know it, you might be calling her name and she might be actually not coming to you right away. Though, you of course do want her to come and you want plenty of love and time together as well! Hope that helps.

24. My child is constantly interrupting and her teachers notice this as well. Why is she doing this and what can I do to stop her? She gets very upset because we get frustrated. Can you help?

    Answer: I can definitely help you on this one. She is most likely unintentionally interrupting, because many people with any autism spectrum disorders have auditory processing problems, or slow brain processing. Therefore, when you are talking, her brain is filtering your words slowly. While you yourself are still talking, her brain has finally "heard" silence, such as a pause in your speech from earlier. Therefore, she thinks it is her time to talk because it is silent. It is her brain that is saying it is silent because of the slow processing issues common in autism. I have always interrupted and my family as well as teachers did not understand this fully either until they were coached on it from professionals. Then, they felt bad and they tried to be more understanding. They will give me a sign to let me know when I can then talk so my brain doesn't have to get confused, although it will and I will start talking during "brain silence." Then, I get the sign and know my brain is behind in the speech so it really isn't silent. This helps eliminate frustration. You can also have a talk with her and let her know what sign she will see from you or the teacher. I hope I have helped and utilize the signing concept. (This is why I am doing this book; to get these important topics out that aren't typically talked about!)

25. I am concerned about my child, her comprehension level, and where it really stands. She seems to know things about what she has read but it seems so distant and not real, almost like photographic. Is there a way to check this out more and how would I do it?

Answer: Any comprehension in autism is good. When you said that she seems to know things about what she has read but the answers seem distant, almost unreal like photographic, I tend to think about her having memorized the book. I tend to wonder, since they are books she is bringing home from school, "Has the class read the stories and been discussing them in the classroom?" So, maybe she is memorizing what the answers are and or only what the words are, and does not really have a true understanding of the meaning behind it. Maybe you can go to the library, or better yet, your daughter's school library with a visitor's pass. Then, choose a book, but a book that she has never read (*the librarian should be able to look up if she has checked it out before*) and you can check with her teacher as well. Then, go home and read it with her. Ask her questions like who, what, when, where and why questions. (*Those WH questions*) The teachers used to do little comprehension questionnaires in the corner of the room or outside of it with each student, including me, and they thought I was doing awesome but all I was doing was basically repeating the book verbatim. They forgot to ask the WH questions as I mentioned above. That is my advice for you and when and if she gets frustrated, let her know its ok to "not know for a while." You can try to explain certain answers by highlighting your voice to kind of show/speak where the answer is. However, do not push to a meltdown if you can help it. Just try and see. I hope that answered your question!

26. What types of things do you repeat and do you know you are repeating them?

Answer: Well, for the first part, I like repeating things like TV commercials; Lerner and Row, Stanley Steamer, things from movies or shows like UP You Go from Monk, etc.! For the second part of the question, do I know when I am repeating something; My family says I repeat things when I DON'T KNOW I'M REPEATING THEM such as when we are walking or in the car, eating, when I'm in bed, in the bathroom, etc. So, no I am definitely not always aware when I am repeating things.

27. My child keeps laughing at things. We will have to watch one little scene that is about thirty seconds long about thirty times, and it is a struggle to get the remote away. Is there any advice you can give and is it solely autism related?

    Answer: Ok, well I have to tell you that when you said your little one will watch a scene that is about thirty seconds long that I can watch things that are literally only two seconds long and keep repeating it with the television remote. My family has struggles with that too (*believe me*) but they also kind of laugh at it. Something that is funny to me for hours and months and years will only be funny to them for maybe a minute or even less. I'll start laughing out of nowhere and they ask why I'm laughing. I start to script or repeat and then they get it. There are some people that say, "Not again." When you asked if it is solely autism related, the extreme extent to which it is being done is pretty much the autism flag waving, however sometimes neurotypicals who are staying up late at night and get weird with their friends can get like that as well. However, since your little one has autism, I would say it is solely autism especially since it's a big struggle with the remote. My family gives you sympathy there. I hope I answered your question to the best of my ability and that you will be satisfied with it.

28. Do you use the advice yourself that you give others?

    Answer: I know it seems kind of funny, but sometimes yes and sometimes no. The reason I say no for some of the advice is not that it does not apply to me, but simply that since I am autistic, I would have to be reminded to use the advice I give to others. This is why there are local support groups for people with autism spectrum disorders & the help of therapists. When I am having issues, my service dog will help me to remember certain things but it is still hard. So, with that, I do remember and use some of the advice I give you when I can remember. And when I cannot, I get the help of others around me who know what advice I have given, therapists, etc.

My parents wanted to read the book to make sure that everything was correct (*although they already know my memory is almost always right*) and to help with some grammatical stuff. However, leaving it to be my "flavor" since I want you all to read how an autistic person might truly write without it being doctored, if you understand what I mean. I left this LAST personal question to put in until after they had read it because I really didn't want them to know my answer about it (*plus I didn't want hugs, ugh*). So, the question and the answer are below:

29. What is the one thing you wish most could happen despite your autism? Or, how does it make you feel?

    Answer: First off, it is that I am on the spectrum and as an adult, so that makes it hard. I am at that legal age, 22, but chronologically on a different age scale in autism. Yet my body is still going through the same spurts that it would if I was neurotypical. I actually cry sometimes a bit because of the changes going on and how I feel very different from everyone else my age. Also, how I just wish for one day; just one day that I could be neurotypical and be able to be like my brothers. Like you know, build a house (*design it*), see what physical love is like in a good way (*just like a hug and kiss- the other stuff, heck no*), taking care of a child (*any type of child*), and being able to cook amongst other things. That is what tugs at my heart most right now, because no one typical can understand how it feels to be right outside this window and not be able to open it, not even for just 24 hours. The reason I really like people to ask me questions on autism is because it touches my heart; it's like I feel that the window I can't open "cracks" a little and I can feel a slight breeze as to what it feels like to be neurotypical. So, I thank those for giving me the opportunity to answer questions. Similarly, it gives me some of the same feeling to write this book for families, doctors, teachers, etc. So, that is my answer. Sure, there are many other things I wish could happen, but that is what I wish the most despite the autism. That is how it feels.

I want to add from my above answer that I would suggest things like Webkinz®, Sims®, etc. because you can design your own house, can take care of your pet, and do the things virtually the adult person with autism wishes they could do if they cannot. (*At least at that time any ways*) It also helps the person with autism learn things, as there will be questions and different scenarios. Therefore, they will be better prepared for what comes along in life. And, they may get a chance they did not think they had through learning it in a virtual world!

I hope that some of my answers have answered a question you may have been wondering about so far.

12

AUTISM AND SCHOOL

This chapter is so important. Visit my website again for videos and pictures to help explain. I will explain later on what school was like for me and what helped. I have additional tips in here, expanded explanations and more examples of my life after the initial tips. I know, I keep explaining this. This is common for me because of the OCD aspect.

Tips for teachers:

Bullying needs to be stopped. Too many times, teachers and others at the school will say, (*Now, this is just some teachers and some people*) "Unless you're bleeding I don't want to hear about it." Or, they will say, "You need to learn how to take care of these things yourself so that you can handle them when you are older." That is not bad, but in retrospect, there are two views as to why I would suggest intervening. You want your students to know that they can go to get help from an adult before a problem escalates too much. If they don't utilize help, then when they are older they could result in taking matters into their own hands inappropriately. For the second reason for intervention is that in later years of life, when there is a conflict, a police officer or hired security person will get involved to mediate adult confrontations. So, someone is usually involved in these matters anyways. It is a good thing for your input as a teacher at these formative learning years while the child's brain is developing in different situations such as cause, effect, respect, etc. A teacher is more than just someone who teaches material, but a person who can help form a life! Alternatively, some teachers will not resort to correcting it right away or properly. Watch the video excerpt on my website. It is what school is like for most people with an autistic disorder and PDD. I have my experiences later on.

My two quotes:

1. With our individual intelligences, we can be of assistance to others, not only to have them benefit from our abilities but also for us to devote our energy to our own eye-catching skills. I trust that if I work with another who has a skill that is not one I electrify, our specialized skills can merge together to create something or make a change! It is never too late to start using our skills, and usually these skills can become desirable to you yourself.
2. To me, "Teaching is sharing parts of the world YOU know about to others who have not yet had your opportunity."

I have tips given below to help teachers understand, work with and use in order to help the student with autism learn and have fun. A student and/or child should have fun when learning.

1. Seating changes are terrible for the child with autism. Do NOT move the child with autism. Also, use priority seating.
2. People with an ASD sometimes get confused with words and speech. It is part of auditory processing disorders associated with all autism.
3. Along with the auditory processing comes asking the same questions over and over, or repeatedly. It is not because they are not listening, but because the first few times, their brain was either not processing it correctly or had been overstimulated by other things going on. Do not get angry, for they cannot help

it. In addition, you might say their name four to five times or more and they may not respond. It is the same thing. Their ears did not process the sounds correctly or quickly enough.

4. People on the autism spectrum can have trouble with reading aloud because they can easily lose their place. Have reading strips available to the students in your classroom. The picture shown (*in my website*) is similar to the reading strips. However, other strips can have smaller print lines with a border so you see only one sentence at a time. You can also have whole page color sheets to put over white paper with black writing, as some people on the spectrum have a hard time. This is because it is hard on the eyes and worsens the comprehension. (*The black print on white paper*)

5. Try to keep down the rowdiness in the classroom, but not through frustration or loud voice because this could upset the autistic student.

6. Use many visuals, especially ones that seem very fun and intriguing.

7. Set up a network for the child with autism/PDD. If you are going into a group setting, have a group set up of kids that like and have fun working with the child with autism. This is so that the child does not sit there or stand there alone while everyone is in groups. In addition, this prevents the child with autism from unfortunately being stuck with someone who would make fun of them or just not encourage them. (*Otherwise, ignore him*)

8. Children with autism can take criticism harshly. Just act as if everything is cool and use markers or pretty colors to put in the right answer and explain...don't use X's or checks to mark wrong answers.

9. Use the (*again check my website*) computer program and recommend it to the parents. Computer programs used for kids with dyslexia, autism, reading comprehension difficulties,

blindness, hearing problems (*auditory*), slow processing, etc. for a small amount of only $90.00 makes a world of difference. It increases enjoyability of reading and learning for kids who start to hate it because of these various difficulties. It really turns them around. Other programs similar to this one cost $800-$2,500 and really do practically the same thing. You cannot go wrong with this program teachers and parents. In my personal experiences section to do with this topic, I am adding my information on hyperlexia, which is similar to reading disorders, but can also be a gift.

10. If an autistic person needs their space, let them have their space. Do not make the situation worse. There should be a room in your school with a swing in it or an autism room to cool down in.

11. Allow bathroom breaks and as getting older, the student should not need to ask to go or sign a signature book as autistic people have lots of yeast and bowel issues as this causes them to be pointed out, etc. and should be an accommodation if needed. Once they have to go, they have to go right away. Also, allow for eating/drinking in class if needed, hence tactile and blood sugar issues that autistics have (*some do*) if it is safe to do so.

12. Clarify directions and maybe even mark down on a piece of paper what they have to do. Sometimes extra time is most definitely needed and warranted. I had an hour a day where I could use that whole time to finish homework that I could not complete the night before because of the autism. I will mention in my personal experiences section an example of why extra time may indeed be warranted. Please keep this in mind as well as having your student do only half the problems assigned. I also explain that later!

13. Note taking is good, especially since handwriting can be literally difficult and can be painful for any person with an autism spectrum disorder or like PDD-NOS. They should NOT get

into trouble for their handwriting that is hard to read or if they take a long time to copy stuff down.

14. Have a private test room if wanted and or needed. – Extended time as well.

15. Many people on the autism spectrum can remember whole movies, but they can forget things like permission slips or even work. They can forget their pencils or literally lose these things along the way.

16. Fire alarms should be addressed. You should have the class or at least your student with autism out into the schoolyard before the alarm goes off.

17. Keep your voice quiet and sometimes robotic or singsong. This was explained in another chapter with how to keep attention and calmness – chapter nine.

18. Sometimes, the child with autism will have to move around outside of the desk and sometimes need to have their chair replaced with a ball chair. They might also need breaks for beanbag chairs in the room. This is because of vestibular and sensory issues that can make the autistic person feel very unbalanced and sometimes even dizzy (*some can get to that point*). Therefore, they need to be able to move around. If your student is tipping back in the chair, they are needing input. I will give you examples and my experiences with this tip as well.

19. In no way should you ever ask any of your students to give out their scores aloud. It is actually against the rules. Many people with autism who are mainstreamed can have major discrepancies in their scores compared to their other classmates without autism or learning disabilities.

20. Phobias directly related just to learning (*PLEASE LOOK AT THE CHAPTER AUTISM & PHOBIAS*)

21. Give transition time, usually about 4 minutes with a visual. My teachers and psychologist always knew it was going to be another minute or two past the 4 minutes.

22. For school assemblies, use appropriate techniques. The child may stay with an aid, psychologist or assistant so that they do not have to endure going through the assembly that would cause anxiety and sensitivity.

23. Sign Language is considered a foreign language and includes facial expressions. I write about this more in the experiences according to these tips. With autism, there are exceptions to this part of the language (*facial expressions*) and points.

24. Physical Education teachers, please look at my experiences and also go to my website to read the write-up I did on a student. I have direction on tips for Physical Education classes. This is because we have problems with processing directions, overstimulation from activities, clumsiness or our low tone, transitioning through activities, interacting with a group and many other reasons.

My experiences according to each of the tips given above are now given below in their chronological order (*ex: For number 1, I will give my experiences that have to do with tip 1 in the tips I stated earlier for teachers. They correspond exactly with the numbers above*).

1. Seating changes for me was horrifying. I would literally refuse to get up or go to the line we would get in to hear our new seats crying. The teachers would realize it wasn't going to be a big deal letting me keep my seat as well as those around me, which I expand on in the "some of my experiences in school" section after these corresponding examples. (*I also go into great detail on this in the chapter, "Autism and Change"*)

2. Questioning whether or not a person with autism truly gets confused by explanations I can testify to through my experiences. I had a teacher that had to explain something to me in

junior high. She got angry with me and snippy because of the number of times she tried to explain it. I didn't understand it, as I was not processing it correctly. The other kids made fun of me and it was a scene. Another teacher in high school broadly accused me, saying that I should have understood the stuff she was talking about even since the third grade. I was humiliated in front of people and made to feel upset/angry because things had to be explained repeatedly. My special education teacher was really nice and when I couldn't understand something, he would actually draw a picture and help me through it. This was even for what other individuals would consider simple three step directions. More often than not, we are confused by what you as a neurotypical would call simple things, yet we can understand things that are more complex. That is just how the brain of an autistic person works. I in addition wanted to mention here that autistic people can get confused on just small words and forget what they mean, even if they have heard a "simple" word for quite a few years. For example, there are times where someone will ask me, "Do you want some water?" I will get confused because my brain forgets the word water. The word is like a foreign language to me, almost like I've never heard it before in my life, so I don't know what it is. Therefore, I cannot respond or I look confused and do not answer right away. It is like people who have to re-learn their words after an accident, but for us, it is very spontaneous. It can happen a lot or it could be months before it happens. Therefore, the person with autism might hear a word they know, but their brain may have to struggle to "remember" it, so give them time. Or, you can remind them with pictures or showing the object if they are younger and have a large problem with processing issues. That is where the CD's for brain processing come in, which I explain in another chapter.

3. You may actually call the student's name several times. This does not mean they are not listening to you, but again, your voice is not feeding through their brain correctly. I would be

completely intrigued in something in the classroom or working so hard, that the teachers would even tap my shoulder and they did not get any reaction whatsoever. We are not trying to be disrespectful, as it is like a person who is deaf getting yelled at louder and louder, yet you of course are not going to get a reaction. I want to add a picture of me when I was completely zoned out and believe me, I was oblivious. I was not even getting a recollection of a flash or sound, not anything at all:

4. I had teachers that told me I was not caring about the class and might as well just skip it, because when they got to me, my processing visually and auditorily had me lost in the reading. I later on acquired these reading strips from a teaching tools store recommended to us by my special education teacher. It really has helped me, although when listening to another person's speech I can't process that quickly enough and what I am seeing through the strip to move the strip downward in time. My teacher, who was a psychologist as well, gave out actual colored papers and that would help assist me with my comprehension. It is alleged that it is a psychology knack and that everyone can use it for comprehension benefits, although that is not entirely true, because it is more for people with autism.

5. The rowdiness in classrooms I know can be hard to control. I had some teachers where I would have to leave the classroom, and finally, others who learned to use other options. They started to use carols, using ok pitch voices and even learned to take a situation elsewhere. Therefore, it was not in front of the rest of the class. They came up with lots of unique ways to help stop the rowdiness without light flickering, any loud voice tactics or even frustration. Some of them asked for help from professionals and psychologists at the school. That is a really good thing for a teacher to do responsibly!

6. Using visuals to show what you mean for everyday teaching is wonderful. It not only makes the learning seem more fun, but since autistic people are visual, it is perfect. I had times where I could not understand a subject at all they were talking about, but then they showed it to us with visuals and we tried it ourselves. It helped me to then understand what I was hearing and finally connect it to real life!

7. Setting up a network for students is so vitally important. When asked to get into groups, I would literally just sit there and no one would come to me or I would stand there still. When I would eventually draw near someone, I would stay quiet and just stare. Many times, people would actually say, "I don't want her in our group," etc. However, there were a couple students that were nice and understood. Sometimes, the teacher did not have to talk with them because they wanted to be with me. One time in science class, when a teacher pulled names out of a bowl for groups, I got stuck with a girl who bullied me. She had people come over and said, "Ask us how we work together as a group." The person would ask and then the girl would say, "See. She can't explain what to do for anything or she just sounds retarded. We need a real person over here." It is absolutely horrible to get stuck with these people. Later on, both she and I talked with a school counselor. Afterwards, she repeated some things aloud in the

class to mock me. I have more examples in my experiences in school that is written after the tips.

8. Teachers, criticism is taken so personally and hard by many autistic people. It is a crush to their problems already and makes them feel literally sick. The red markers of check marks made me so angry that I used to hit myself in the face. I even ripped my stuff up if it was even just one or two answers wrong. The psychologist in high school explained autism and criticism to the teachers so things changed. My teachers acted like it was nothing, just put the right answer in nice colors, and then put a smiley face. It is the look of things as well that can really calm the mind, hence color therapy as well. This is so vitally important for the mind of the autistic person and for no outbursts. It will take time, but those X's and red marks is just a disaster for many on the spectrum. Also, many on the spectrum cannot handle seeing the correct number out of the entire number they could have gotten. My teachers started to just put what I got right with an explanation point and a smiley face. Boy, what a difference a small thing can do for the person with autism where mistakes are seen as horrific.

9. The reading program mentioned in my website changed things for me so much. I sought out this exact program after reading evaluations by a psychologist were done on me which concluded that I can read, but I cannot read. To explain, I can see a word but when they are put together like a string, I do not understand what they mean and they can truly jump around on the page similar to dyslexia. I would take a quiz and fail it or do very poorly in classes. The teachers or another student would read it to me after about 10 minutes of getting me to calm down and I would pass it with excellence. In school, especially elementary school, the teachers would tell the students to read an amount in the book and we would discuss it later. I was actually crying really bad and embarrassed in a way, so I would put the book up all the way to my face. They didn't know what was really going on with the book up to

my face and thought I couldn't see the text. It is so hard to describe what someone feels, especially if you get stuck when the teacher comes to you to ask you your opinion on text. If this happens, then the teacher should know ahead of time about the comprehension deficit. I expand on this part in the experiences in school section later on in this chapter.

Now I will explain hyperlexia to you all. Hyperlexia is almost a gift in a way because children especially at a young age can see a word, pronounce it and put it together; almost any word. It does not always mean total comprehension, but rather being able to concentrate, have complete understanding of the root of a word and to outrival it. I believe I have this as well, as the symptoms fit in with me despite my obvious deficit in comprehension of many words together. (*As I said, it does not always mean comprehension of words stringed together, but rather the complete understanding and pronunciation of almost any one complete word, especially when young*). Another beneficial thing about hyperlexia is that it can become an engager and an encourager for speech! Putting letters together is great too and can be fun. This can help bring about hobbies such as word tile games. As they see the board and tiles, that part of the brain goes right to thinking of words! It has been believed that only a handful of people with autism actually have hyperlexia.

10. I had a place in school to go calm down in so overstimulation would go away. I had a place with a swing that would help and I would also go to the psychologist's area. He had a fountain and also a calculator. When you would press the edge of the calculator, it would come up. I would close and press the button repeatedly until I was calm enough to go back into the classroom. Every school should have this setup and even a place in their own classroom if the class has a child with autism in it.

11. I had a teacher that was so angry because I would leave for bathroom breaks. I also was late sometimes because I was

slower. This is because of the autism and the bowel issues that people with autism have. I had to embarrassingly go through a meeting with the teacher and principal to get a "special bathroom pass." This was so that it was a contract to let me go to the bathroom when needed and so that I would not get swept. Being swept is a whole other subject but I am going to mention it here. The officers knew about autism. Therefore, they didn't sweep me for being a minute late to class sometimes because of transitioning issues, bowel concerns and generally being slow because of brain differences with autism. I had once accidentally been put into sweep and started freaking out because it was changing my schedule and I looked at it as a really bad thing. The sweep person there beckoned the security because they found my reaction odd. The officer explained that I shouldn't be in there due to my disability and being a little slower than others at times because of it. I just thank God I was sent back to class where the officer I liked so much helped me calm down.

12. I would spend hours doing homework. My mom would see other children outside playing while I was inside struggling and missing out on important time. I want to add here, that for people with autism, school is no doubt overstimulating and the brain is working harder for those hours. So, when they get home, it is like the brain is so full it can't afford to take anything more in. A break is deeply needed for quite a while. I try to explain like this: If you have a plug on a hose, for quite a while the water will continue to stay inside the tubing. However, after so long, the water pressure in the hose becomes too much and the casing that was used to keep the water from coming out explodes off the hose end. The water squirts hard and fast, equaling the amount of time it stayed in the hose when it was supposed to be going out. Similarly, the brain can't take any more and so lots of homework (*which can be modified, as I had teachers that had me only do half the problems, etc.*) can cause meltdowns. Then, the person with autism is just worse the next day too. Homework modification

should be in an IEP and even in a 504 plan (*if for some reason the student starts out with that*) for many people with autism. Our brain will not only process slower but will also get stuck on patterns on papers and writing. The person should be allowed to finish, especially if it's in school, at a time where they won't be missing out on things like recesses or special activities. I had been having that happen until the ending of junior high school. It will only make things worse and our brain needs time to relax as you will understand more what it feels like to be autistic in an upcoming chapter, especially in a school environment. I also want to add that I was helping in a classroom where I thought a boy certainly had Asperger's Syndrome. The teacher was going to have him sit in the desk and keep on doing the work while he would miss out on a very special presentation about tornado activity. Bottles hooked together that made a funnel with glitter, etc. I told her the situation and why, after observing him, he should not miss out on it. It would only upset him and he was not going to get very many more math problems done even during that time. She stopped to think about it and then he was able to join. This stopped a distress and when he was ready, he was able to finish more of the math problems. It took him a long time, but the work was very good and paced out correctly. I also have typed up what is an emotional, physical and neurological observation I did on him that is posted on my website.

13. I talk about the handwriting problems and issues I encountered a little later in this chapter. Also, it is a motor problem as well as a processing and visual problem. Most autistic people cannot write, listen and see all at the same time. It just is usually not very possible with the feedback our brains have to deal with every second of the day. I had teachers writing things for me and later got a student with carbon paper. It just depended and sometimes I got both. Some people with autism can use a tape recorder or even better, be able to have the same class two times a day that would be put in an IEP. I want to mention that if the handwriting is slowing down class activities that the

person be allowed a special type of laptop that they can use (*I used one at times while in college with another student*). I want to mention here as well, since it has to do with timing, that some activities be possibly lessened. OCD can play into this as well, and if the person is taking a long time to write, they can be given the teacher's questions for the next day ahead of time to do during another hour (*just an hour full of therapy and catching up on homework*). The teacher may not feel like doing that, but if it is in the best interest of the person with autism and the rest of the class, it should be warranted and put down in an IEP. Some of these things seem pretty easy after hearing them and the people should accept them, but people seem to be worried so much about character and how things look rather than forgetting about the image and focusing on the best interest. I wish sometimes that our school systems could learn from some of the countries in our world that are struggling but make school fun. They crave school because they do not feel so pressured for image, but because for them, it is a blessed opportunity. They feel great compassion that what they learn in their pace and time will convey to their villages! If a person has to verbalize something, they may feel better with it written down in front of them so their nerves don't overwhelm them, especially if they are worried about points. Again, I think people should focus on learning it uniquely and not be judged so much, as some learning curves here in the always growing US do. These are the suggestions I have in this area.

14. The tip pretty much says it all (*separate testing room*) but I want to add that it is because of nervousness (*along with that, correction worry*) and sounds of other pencils, clocks ticking, etc. Sometimes my college teacher would unplug the time clock from the wall because with my sensitive ears, I could hear it. I would be doing a small quiz or something but become distracted. Later, shades were also hung up in front of the blinds. He was a teacher who helped bring many accommodations appropriate for my disability to light. He was awesome! ☺

15. I cannot tell you how many times I would lose permission slips or pencils, even homework, etc. The teachers finally learned that the best thing to do was to call my parents and let them know what would be coming home with me. Since my parents would pick me up from school, they were actually handed the things sometimes. It is hard to explain how we can lose such things but it also has to do with getting distracted from our senses, motor problems as well as cause/effect. It does not do any good to yell or get angry with the kid when this happens, as it is usually a part of the disability most of the time. Also, similar in nature to this, I had people who would tell me to call them or meet them at a certain place. I would not show up so they were many times offended because they did not understand that I had autism and that was why. However, they started to understand more as I got older. I still have to tell people that it just happens and I am sorry. The visual and time schedule really does help with that though. However, sometimes I actually have to be reminded to look at the schedule even if it is with me all the time.

16. Fire alarms are so difficult because of how loud they are and the mass amounts of people leaving the premises. I explain more about the fire alarms and that situation in the first chapter. If you missed it, just go back and skim through to read it.

17. This tip (*voice inflections*) I explain in the chapter about doctors, dentists, etc. You will have to read that in order to understand how to use it more accurately. My teacher had a unique voice and exaggerated it with great inflection to help me. I would hear neurotypical students saying, "You'll never fall asleep, misunderstand or get bored in his class. His voice is so interesting and very helpful in so many ways. It is actually a class that makes school something to look forward to." Also, a good representation of a type of this is in the movie, Silent Fall during the time he is trying to get the kid to pay attention. I have a video clip of this on my site so that it will be clearer to you.

18. Especially since I have gotten sick, my vestibular has gotten a little more out of balance. I have had to quite a few times sit at the couch or on the floor during dinner time and other meal or snack times. It causes unbalance and can sometimes even cause dizziness. When I was in school, I had to get up a lot. The teachers actually didn't mind and they started to understand. It was very helpful and after just a small bit of time, I was able to go back to sit down. You will also see a lot of kids sit on their legs for input and support, but you don't want to let it go on too long because of the flow system in our bodies. A kid in high school who was really neat and cool, figured out that I would pay attention more and feel better if my chair was either tipped backwards or with the back end up in the air. He was seated next to me, in the back (*also because he thought I was neat too and it was arranged with the teacher, meaning that she saw it as good when he asked her and explained, etc.*) and he would with his legs lift up the back end of my desk and sometimes bounce it. He said it was a good work out too, lol, but really cared about me and knew that it made a difference. So, he decided he would help me out. What a nice guy, ha? It really does help and these all have to do about autism and the vestibular system, along with autism and the sensory system.

19. I would get embarrassed and sometimes laughed at when I was asked by the teachers to give my scores out loud. The other students gave their scores aloud too so the teacher could write it in her score book. In high school and even before that, the teachers started to realize it upset me and would ask my score only if they knew I had a perfect score. This was so it would help me out because the other kids would actually then talk to me and smile. However, in high school, there was a really mean teacher who I had for two classes of the day that would ask everyone their scores aloud (*it was a special education class but most the people in the class were higher functioning than me and in academics as well*). Lots of times I would only have about 7/29 or 30 right. She even mocked me and said I should have

learned this in the 3rd grade, etc. (*Sometimes this was because it was a review of typical English terms or mathematical skills, etc.*) She even asked the person behind me (*who was really nice to me and we all made jokes together about her in sign language to make me feel better*) to teach me how to understand the answers. She kept doing this and mocked a girl on her spelling test, accusing her of not studying, but she had a form of dyslexia. The girl did not care though and just made herself feel better by our whole classes knowing how mean she was and making fun in sign language. Unfortunately, this teacher is still teaching. But, at least now, she had to sign a contract with the administrator that she would stop asking grades aloud. People with autism usually have deficits in at least one area and are upset by lower scores. They should not be made to express them in front of others who wouldn't understand why they would get such low scores, even if they did understand.

20. Look at the chapter "Autism and phobias"

21. Transition time is very crucial at times because the autistic person's brain is very set on one thing, almost like a one track mind. Most of my teachers actually had me transition earlier than the rest of the students to allow for better preparation. I would also have a visual to show me how much time. The psychologist suggested another 1 to 2 minutes would help, which it did. It can be very anxiety provoking for the autistic child when going from one activity or class to another. Transitioning is also part of letting the person with autism know if there is going to be a change. The psychologist would go over things several times with me so I could be prepared, and even then, it can be hard still. At one time, the psychologist had wanted me to come in for some testing but I was not told of this. So, when he came and got me (*he knew I was going to be upset*) I was very distraught. He ended up calming me down and then eliminating the testing to another day when I was prepared. I also want to mention here that I would have band class where the room would be changed or turned around the other way

for directing. I would be so angry and upset by the change, so the one director would sit next to me at that time for a bit. His charm always calmed me down… he was the greatest. He would also tap me on the head with the directing stick sometimes, lol. (*Of course, there were other great people, but I just really loved this director*)

22. I have mentioned the subject assemblies earlier and will later in this chapter mention more about this issue.

23. I refer to more deeply in the next chapter about facial expressions in sign language and the difficulty it plays on the autistic being because of our limited ability of social awareness.

I really wanted to give examples so that you are not just receiving tips, but also have a situation to think over so that the tips can be utilized more efficiently through learning.

Great ways for the autistic person to learn: (*different styles*)

1. Mentioned before was visuals

2. One on one instruction (*through aids*)

3. Trying things hands on

4. Learning through music; a song so can make more sense

5. Tapping – this is a unique method of helping learn. As you are learning a word phrase, you tap your hand so that it helps the nerves in the hands connect to the brain, causing it to process more efficiently.

6. Repetition – I cannot express how much this can make a difference. At first, it can be confusing or stressful, but with autism can be great in the end, since repeating things is commonly efficient as well as noticeable in autism. For EX: You have a card with a symbol such as the Pie symbol. You place the card in front of the autistic person and have them stare at

> it for a few seconds and then say "Pie." You remove the card and put it back (*You do not have to do this after about the 10th time*) and the person says "Pie." Do this several times. Then you would go on by showing the "Pie" card and add a card lined up next to it that says or has the symbol =. You repeat equals several times. Then you put them together and repeat several times, PIE =. (*Like PIE = 3.14*) This really engrains it in the head and stores it in a precious part of the autistic person's brain, the part that see's something or hears it and can then repeat it back. This can be utilized with *and for* almost everything. It can start to be fun.

I also want to mention that for a person with autism, it takes their brain longer to pull up an answer or a remembered fact. This is because the tube in the brain actually acts like a leak. It is supposed to flow through smoothly, but instead drips, causing longer time to get that part of the brain stimulated. This is very important for teachers and parents to know because it is scientific and gives real explanations so that there is no questioning. It is so very important. I think as you can tell, every chapter in this book is very passionate to me and I hope that will rub off to those who are reading it. I had teachers who thought I was being slow on purpose when I would take a long time on tests and then beg for a bathroom break (*they would say I could not leave during a test but an accommodation that can be made*). They did not understand this actual scientific explanation of slow brain processing. Many teachers will sometimes play games where the person goes around and repeats the people's names in the circle. It can be upsetting because of this struggle in autism and then result in teasing, etc. When doing that game, have the autistic student be the first student in the circle so they only have to say their name.

In addition to this, as it is similar in nature to the subject at hand, I want to add something here quick that I think should be brought to the attention of all automatic answering systems (*and that goes for school answering systems with choices – Ex, for lunch room press 1, etc.*). Many people with autism can have a hard time with these automatic answering systems and choices they give like medicine refills, doctor clinics, companies, schools, etc. I think that it would be helpful to the people of the community who have this slow brain processing (*for some it is not autism but a result of war, stroke, or other disorder*) to in the beginning of automatic answering systems, have the choice to pick a speed at which the system talks. I also think as a plus, there could be a number to press just for people with disabilities who can talk with a person and not have to keep going over and over through the choices when they get lost and confused. I just had to bring this up. It does involve many school answering systems, so maybe your schools can have a meeting to think about that!

I will list below some of my experiences in school. Most of them were bad, but some were absolutely wonderful! I break the two categories up.

Before going to those categories, I would really like to bring up a couple concerns that parents do deal with many times in schools and that the teachers have to handle as well. I want to bring light to these subjects before going into my own not-so-good and good experiences in school.

1. Correction
2. Corrals
3. Little walls
4. Conferences with teachers

I will start out with Correction as it can start to become an everyday occurrence for some kids and causes teasing, bullying, etc. It also causes upset for the kids with an ASD because it is going into the literal part of the mind. Some kids with ASD get upset when other peers/kids are not following the rules. He/She can get upset that the others are not following the rules because it kind of goes along the lines of "taking things literally." He/She is literal in the terms of "This is what you can and cannot do." Therefore, He/She is adamant that the others follow this too. He doesn't understand why the others don't understand what he knows, because in his mind, everyone is like him. I used to get upset myself if others didn't follow the rules, and believe me, it caused havoc for me sometimes. But, the teachers and others were glad that I was their little "watch-girl!"

I will now go on to Corrals. Sometimes I get words mixed up that sound the same but mean different things, so of course someone says corral and I think of "riding a horse in a corral." However, I now am also thinking of "school corrals" because I have used them and know what they are, give advice, etc. For the paying attention with the different smells, noises, etc., and because everything is coming into the brain at almost the same level, concentration at many times for a person with an ASD is out of the picture. In our room, we had corrals and sometimes a tent over them. This was a special education room though, but even some of the regular education rooms had them as well. It helped to block out the other people and the other sites in the classroom. A good tip I have would be to have the school psychologist assign the child an aid as well to help them throughout their day. It can make a world of difference (*it did for me!*) and somehow the class seems to behave differently then, kind of like when drivers see a cop car they straighten themselves up. LOL!

I will now go on to Little Walls. Little Walls is a term for having a small divider the person with an ASD can have to go behind so they have their own space within the classroom (*which also gives the option to stay within that space without leaving to a calming down room*). Sometimes, parents will choose to have their child assessed for developmental disorders. If there are Little Walls used by your child in their classroom, then it may be an indicator that the school psych. staff is investigating what disability your child may have before a full assessment. That happens a lot, depending of course on the different school districts and of course the various special education departments in your very own state.

Certainly not last and not least, are teacher conferences. Sometimes, and many times, let's face the truth, we have to sit down with the teacher to try and figure out some better methods, even between IEP meetings. It can be stressful for both the parents and teacher as well. Some parents get so angry with the regular education teachers who have their special child mainstreamed because they can seem sometimes so unwilling or so quick to put off the meetings. Other times they seem hesitant during a meeting and act stressed, making you as the parent, feel like they don't want to take the time. However, I am here to tell you what may be behind some of these behaviors. One, teachers get nervous because not only do they have to fit in all the curriculum and pre-tests (*which are so many it is making the special education department and regular departments stressed and literally sick*), but then they have to modify work and curriculum according to the student's needs with autism. Then, they have to try and manage the behavior of the child with autism and the other kids at the same time. (*Teacher aids I think should be assigned to classes in almost every class through junior high because if you look at it truly from any view, a teacher can't possibly handle every situation while taking care of every student in a full classroom by themselves, especially when they are youngsters*) They are so monitored by their curriculum standards, the special education department, their boss and others. So, when there is a meeting, the groups the teacher is looked over by may be supportive or in some cases, unsupportive. This causes great anxiety when conferences are documented and reviewed. So when the teacher's social skills seem compromised or may seem intentionally negative, therefore making you feel bad, think about the great possibility that they are just worried of being attacked and fitting everything in correctly and safely. Try to be very calm and even write a letter at the end thanking them for the conversation so that they feel more comfort. It will help them with the groups surrounding them. They (*the groups*) may see that small note you gave them (*the teacher*) and then try to see that the teacher is being responsible. It will lessen your anxiety and obviously the teachers as well, resulting in better communication, which leads to better ideas. They should know (*the groups*) that meetings do have to be called and it's the responsible thing if you have a child with any challenge, but they don't always see it that way and jump to conclusions. This is because those groups have groups watching over them as well. Therefore, that is a small thing you can do to enhance these meetings into being more positive for you the parent, the teacher and even the groups overlooking the teacher. It should not have to be that way, but order of rank always gets higher and the power seems more intimidating for those one rank below any other.

I hope I have enlightened those four topics that are related to school many times. Now I will below go on to my not-so good and good experiences in school.

I have here a short synopsis, but a very good one with details below.

I did attend public school. I was in both special education and mainstream classes. I could not handle eating in the regular lunchroom and to tell you the truth, school was a severe nightmare for me. People would drop books just to make me start screaming and holding my ears, moving things on me so I would start flipping out and stim with my voice, etc. During the assemblies, I buried my head into my lap and couldn't handle that. As I got older, I stayed with the psychologist or counselor during those times and they had a schedule to help calm me down with the change and everything. I would take off my shoes and socks to stop the bad feeling, pour water on my desk and lay my head in it, line things up, get into trouble for sloppy handwriting and being slow, make noises and be told to be quiet, get lost in the reading, and I should have been diagnosed in the first grade. However, the teachers and schools I went to were possibly too afraid to say anything until I got into the second half of my education. The abuse from others was more than anyone could handle. The police at the school had to get involved in a few cases as I got much older and started to verbalize some of the things that were happening. I started to bang the walls and lockers in the band room where I went to vent. My band director would calm me down and sing or play an instrument for me. I would kick the garbage cans and have meltdowns (*especially from a special education teacher who should not have been able to be one*). The psychologist or counselor would come and get me out of the situation. Honestly, a few of my teachers were frankly awful. Physical education was a nightmare too. I would not understand directions and would always be the one that everyone said, "No, I don't want to be next to her." It was horrible, but then I met a girl and she had special needs too, so we always had to stick up for each other and even pretend just to get through the day. The teacher was just terrible too unfortunately. However, Physical Education teachers that are understanding and willing to learn about autism disorders can be a shining light for the autistic person and can also help bring confidence. I also met another P.E. teacher when I was doing a write-up on a student I mentioned earlier and he was absolutely awesome! In high school, the psychologist said I could be exempt from P.E. or I could do it correspondently (*on my own*). Some other classes I was exempt from as well and I got lower standards for those new tests taken to graduate. I want to mention that there can also be another way besides mainstream or exemption such as adaptive P.E. I want to mention here that most states have the standard that the person with autism is allowed until age 21 (*public school years*) to be in school. To be kind and fair to the autistic student, the student should be able to walk and get a replica diploma with their friends and graduate class but then have the option to go back for another couple years. Many students won't want to stay, but some it can be good for. It is just more fair the way I stated above and you really should speak with an autism or disabilities advocate if the school is saying your child cannot walk with their class they would normally graduate in if they want to come back. It is a good law to help these students, but isn't good if it threatens or upsets the student and friends, family, etc.

In elementary school, kids would do everything that they knew they could for every sensory overload they could. I was dragged across a field and thrown dirt clumps at during recess when I was trying to stare at the bark on a tree. The girls had trapped me on the bus and in the bathroom so they could take my lunch card. I would then get in trouble for "misplacing" it. I was told to mimic things they were doing/saying that I did not know were inappropriate. EX: My teacher had a miscarriage, and a mean girl told me to go up to her to make her feel better and say, "Sorry you lost your baby, B****." I did not know it was inappropriate or hurtful, as I was mimicking. Luckily, the teacher knew I would never talk that way. There was other abuse as well in the mimicking area, but I do not want to go into details, as it is very personal and traumatic. They also would tell me that they were going to kill my family and me if I said anything, which I then did not for a very long time. (*Years*) They used to say how my cat, who really saved my life for all those years of school, was going to be dead when I got home and how her head was going to be in the freezer. They would also make loud noises, change my environment; you name it, they did it. In high school, girls pointed me out to people who walked by and said, "Look at that girl sitting over there… she's retarded." Again, they would make it so I would mimic or echolalia sometimes things that were inappropriate so I would get into trouble. My parents never knew this, because I kept quiet, and to be completely honest, I was relieved to be home; away from the millions of people, noises and smells. In the P.E. room in junior high school, I was sprayed with four different perfumes, which made me gag and throw up into the sink. The girls used to always do things and the boys as well. I would try to join a group and they would say, "We don't let retards in our group." They would hide my work and tape it under the desk, etc. I had picked up a chair when my teacher had left the classroom to get something, because I wanted to throw it at them, but I did not get the chance to. Because of the frustration, I would grit my teeth so hard that it felt painful and like it would break them. However, I did end up really breaking my eyeglasses and needing a new pair. School for me was a nightmare and people would not listen to me. You will not always know your child is being abused, but if they are on the autism spectrum and mainstreamed, most likely they are in some way or another. It needs to be stopped. I went to a 1st grade classroom where a young boy I was so completely sure in my mind had Asperger's. I explained to these kids in his class both individually and in small groups at recess (*not using the word autism, nor pointing the boy out*) lessons in "Don't make fun… you don't know the reason behind the behavior," etc. I am going to write a manual for all teachers on this and it will be a laminated lesson plan for them some time after this book is complete.

I think all teachers need to do this, starting at a young age.

So, time to hear some good experiences…

I explained in my personal questions about the Police officers being so kind and the psychologists, etc. That was great. I was also in band and we went on a couple trips. The assistant band director was the greatest guy I know. He kept me on track, helped me, would let me miss things that would overwhelm me like playing at an assembly, etc. I had been really angry once that I had to give him my key for the locker because I kept losing it and then finding it. I was so frustrated and upset that I threw it in his hand when my mind was not handling change. He actually did not let go of my hand but just patted me on the back (*hard*), gave me a hug (*tight*) and told me I was going to be ok. He, I think was my angel. I made a really special plate for him that took me days to make at a pottery place and I put "My Hero." He would let me rearrange and fix his key holder for all the lockers and let me stay in his office with him into lunch hour, etc. He would always help me out when I was upset and did so much for me. I had a science teacher amongst other teachers who would talk to me about stuff I was really interested in and even joke about the people who were not nice to me. They were always on my side. Then, there was the front office where I would stamp things for them and had food with them every day after school. There were some great people I met and we are still in contact all the time. The ESAP class was wonderful and we became like a family. We helped each other out and we were very similar to each other, so we met all the time and would go outside and have fun. It was wonderful that part as well. Those were the wonderful experiences and they did come late, but were wonderful and really helped keep me going.

I want to mention something neat is that some of those teachers that really wanted to help me out ended up at the college I went to for a while to be a teacher there (*lol, they still kept an eye on me*). A couple of the others still worked at the high school and couldn't leave to come to where I was during the day so ended up joining a community class to see me!

Yes, with autism, school can be negative. There really can at least be some wonderful experiences.

I give this advice: Be in contact with the psychologist, the teachers and the aids at least once a week to be on top of things. This is so very important.

Also, aids can help because it's like the teachers and students have to behave when they are around. Talk to your child about this option.

I hope you have read and learned lots from this chapter, as this is really why I wrote the book; I just want to help people and talk about some things that aren't being talked about in regards to autism spectrum and related disorders.

13

College and Autism

My accommodations are listed below that I had when I tried some regular college.

Note taker

Priority seating

Tape recorder

Extended test time

Private test room

Extended time on work when called for

Books on tape and reading system

Reader for tests

Clarification of and for directions and assignments, etc.

Reduced class-load & insurance waiver

Also:

Being able to fidget with some objects, sometimes eating because of autism and low blood sugar/sensory and bathroom breaks even during tests

Foreign Language and facial expressions

I was in college for a little while. There were many moms at community college and smaller classes, so the people in the classes most the time were really nice unlike the other school years. However, teachers may still point out, be unkind and try to refuse accommodations; again <u>some, not all</u>. There are students that can also be rude at this level.

I was taking a sign language class (*because I already knew sign language*) that was fun and helped to encourage interaction because I could help the other students in the class. I was getting lower grades and the teacher said, "You are not utilizing the facial expressions so tons of points are being taken off." (*This had been happening in high school as well, but the teacher realized it was the autism, as she worked with many special needs kids and decided to eliminate that part of it, or in other words, decided not to punish me and even have points in that category for the class*) I went to one of my favorite teachers who also happened to know about autism. He said that because of the autism, it should be an accommodation that facial expression points in the class be "in words" thrown out. This is because this is part of having autism and knowing the language is what the class is really about anyways. So, I then explained that to the sign language teacher and it helped him and I. When things in the room would change, I would have horrible expressions or panicked expressions, so the teacher then would joke with me. He would sign, "If you could only naturally use expressions like now you'd be more effective!" He took that other teachers understanding and decided that because of the autism and trouble with expression, it was not worth it to have it be worth points. It ended up fine in the end and he tried on the side to help me with facial expressions by showing me pictures. Well, it did not work very well or I just looked so fake!

Please bring a paper on students, teachers and autism. Lots of the same ideas work as for the other school years, so you can print off my page on teachers and autism along with this page as well. You can get it off my website or highlight these things in the book, print and hand to the teachers. I encourage you to highlight and hand the information here to the teachers since the tips in this book are better expanded upon. Therefore, it would help them by giving them real life situations to compare to. I cannot wait until this book gets more into circulation, because it will be sent to school districts I hope!

As far as what you will do with DRS, also referred to as Disability Resource Services, is explained next. You will get a form that is referred to as an INF, which states your accommodations that your college teachers are supposed to provide. It is like they are signing a contract to do those things, so if they are not, they will have to meet with the DRS and abide. It is also good to ask the DRS person in charge what teachers they would recommend for you for your specific disability. The person who is meeting with you usually likes to ask how your disability affects you. However, if they are getting to ask too many questions and making you feel uncomfortable, let them know. You can print off some of the accommodations from my website under Autism and College and just hand it to them. You can also request that if possible, there be an area in the Disability office that you can go to calm down. I used to have a room in the DRS office and when upset or something I would crawl underneath the desk and calm down. Many DRS offices have counselors and yes, even sometimes a psychologist. If they are pressuring you to see a counselor, this may be because the DRS is paid to have these counselors and if they are not being utilized enough, they can stop working for them. In addition, getting in touch with a psychology teacher at the school can be helpful. At times, one of the psychology teachers is also involved with the DRS office. I personally wish at regular colleges and universities, an autism room would be dedicated to students with autism so they can cool down in it, socialize and teach others in their school and community about autism! How awesome would that be?!

I had a problem with having a reader, even after several teachers tested the theory out. After my test scores went from failing to passing within a few minutes of having the teacher themselves or another student, note-taker, etc. actually read me the test, the DRS still would not permit me to have official reader accommodations for any tests. The psychologist there even saw this with me and the DRS person in charge of this type of accommodation would not listen because they said I got "ok" grades and don't have dyslexia. These teachers and others who knew about autism and learning disabilities tried to explain that comprehension and slow brain processing should qualify the person for a reader. I was taking a test and started crying really bad because you had to read a page and then fill out answers based on the reading in the subject. My teacher came in, read some of it for me, and then I started to get the answers right (*which were completely making no sense beforehand when I tried to read it. He knew that I knew the information and he was a psychologist at the same time*). He finally after a few minutes of trying to calm me down and reading to me, mentioned that I had an A in the class and I did not need to worry about this. I needed help with the reading and there wasn't a system to help me there. That made me feel better, along with the teachers or students who read to me and helped me understand the text during regular quiz days! Comprehension issues of reading could give an automatic "reading disability" to autism, which would then have the Reading System available on INF forms immediately. Some autistic people are fortunate to comprehend but most will need a reading program and pictures sometimes even too. (*This is extremely rare, but a couple in the world have a special savant ability that allows them to read both pages at the same time with complete comprehension*) The teachers and student helpers used this with me along with visuals. Also, the DRS person cannot deny a reader if the teacher chooses to read the test to you or have a classmate read it to you. This is because it is "their" classroom and "their" decision even if it is not an approved accommodation.

I really would like to mention that if your child with autism wants to try college, Voc. Rehab should let you try it if you really want and not just automatically say that you have to find work with them.

Though, I must say, they can be good and unique. Because of autism and especially the change in how things work from the school required years and college years, the person with autism should start out with less classes than a full load. They should start with classes that may be easier for them, maybe even a little fun! People say that junior college is about two years or a little more before you graduate, but it is certainly not that way for everyone. For the autistic person, many of times the insurance says that they have to be taking a "full class load" to still be covered under insurance during college. However, because of autism and its complexity, you can get a class load waiver that your psychologist, psychiatrist, or doctor signs that then is returned to the DRS (*saying you do not have to take a full class load because of the disability*). Then, it is turned into the insurance. When it is turned into the insurance company, they are then aware that your child going to college has the right to be covered without a full class load because of their disability. I would also like to mention here that some insurances (*under ADA and their obligation*) can keep the person with autism on your (*the parent's*) coverage as long as you have it no matter what age they are (*or until they have their own*). This is aside from insurance that helps with covering costs for meds, etc. that is a cost containment system for the person. That insurance I just talked about would then be secondary insurance if still under the parent's insurance.

If I get better from being sick and get past the PTSD from it, I personally would maybe take only a few classes at a regular college and then go on to a specialized college for people with autism disorders. The states are just starting to create these great schools. I would go there so I could learn with people who understand and accommodate to your needs, be with others like myself, and be able to learn these great things while having sensory, living skills and even psychologists incorporated as well!

14

AUTISM AND COPING METHODS

Autism, since it is such a complex disorder, causes great stress because of the vestibular system (*mentioned in the next chapter*), all of your five senses, dental and gut problems, speech and language problems including slow brain processing, etc. There are ways to cope along with therapy. I have listed some of the coping methods that I, myself, have used and that others I know have used.

Ways to cope:

1. Spinning (*such as body, swing, toys*)
2. Holding and chewing on straws
3. Pet and aquatics therapy (*including dolphins and horseback riding therapy aka hippotherapy*)
4. Water therapy (*watching, pouring, splash, listening*) – with cups, hoses, etc.
5. Lining things up**
6. Stimming (*Chapter 4*)
7. Prismatic glasses
8. Watching trains
9. Fidgets

10. Autism therapies (*Mentioned more in the chapter right after Vestibular*)

11. Swinging of the arms (*what you can do if it's disruptive, like in class*)

    ** Lining things up: When an autistic person is lining things up, part of the reason is because the world is as chaotic as it is, that at least a row shows some value of control. It brings order to such an unorganized world. It also has to do with the shape of it (*This will be explained in more detail in the chapter "Autism and Phobias"*). Many times, especially in the younger children with autism, if you move something out of their line you will hear a scream. (*Sounds like eh.eh.eh.eh.eh... until you place it back in place*) You can view a great representation of this in the movie, House of Cards. Other times, they would just get upset. I remember my parents moving a bear that was pink out of a line of my animals that I had around the edge of everything in my room. I started to throw a small tantrum and scream. The psychologists use this method to check off signs of autism as well as to reach attention.

Many people with autism, you will look at their straw in their drink and ask, "How can you even get any soda through that bent straw?" This is just one example that many autistic people chew on straws as part of a way to relax in restaurants and also for mouth input, as well as jaw input. This is actually a good thing because it is helping strengthen their jaw muscles. I have always bent my straw so much that I would only get small amounts of soda into my mouth. I would also carry it around and bite on it quite a bit of the time that passed because it really helped me a lot. If you go to my site, you will see a picture of a small child in my slideshow and coping mechanisms page that is chewing on a straw. That child is not autistic but it is just a representation.

Horseback riding is great for the vestibular system, as it is training the body to use every angled bounce to its advantage. I have been on a horse and wanted to keep getting on the horse repeatedly. It is also the bouncing feeling during galloping that many autistic children as well as adults love! However, sometimes it can take a while to adjust to the new vestibular movement, in which starting out with some swinging and bouncing at home would not be bad for a while first. The connection an autistic person can have with a horse is like no other! Personally, sometimes the movement makes me laugh. ☺

Water can be an automatic love to autistic people as well as dolphins, because once again, it is pure nature. After a while, they can really start to flourish (*the autistic person*) and connect as well as they do with the horses and dolphins during aquatic therapy. There are so many things similar in nature to them and us. It is great to ride on them in a pool, especially because of the flow at an aquatics therapy place that is usually indoors! The dolphin's special noises they make and their unique texture can also help to reconnect the senses.

Prismatic glasses are slightly tinted glasses used for light sensitivity. One of my great teachers I had while I tried college for a while provided different colored paper with black text (*pink and blue*), which helped my comprehension a little and took the strain off of my eyes. It helps compensate for light sensitivity. In addition, a screen sheet might be placed over the computer screen to help with the "energy waves" that an autistic person can see but a neurotypical person would only be able to see when viewing from a camera. It usually is not as strong as that example or else the autistic person could not look at the computer at all. It is the perception and division of color and wavelengths that is so different from autism to a brain developed to what society would call "normal." Sometimes I could not look at a computer screen for very long, so it would become very difficult to do things on the computer and that is why my eyes would squint. In the computer lab at school, the teachers and librarians would be frustrated with me for trying to sometimes stop doing the computer during class. Little did they know, autism has effects of heightened sensitivities for wavelength perception or prism perception. These things stated above (*prism glasses/diff. sheets*) can really help, as it helps for me and other autistic people I know.

When I was younger, I would literally spend myriad time pouring water in and out of cups, watching water go down the drain, etc. It is a calming and therapeutic experience, not to mention water is known to some religions and cultures as very healing and spiritual. It is no wonder it is a miracle worker for many people with autism. I personally have a water therapy CD that I listen to which works great. It feels so refreshing to the senses. I will still get fixed on "washing my dishes" because I will be stuck there for a very long time. Water tables are wonderful, and especially when they have little slides to watch the water flow down and wheels to spin the water. It is mechanical-like as well. You may notice that you will have trouble getting your child to get into your bathtub (*transitional difficulty*) and in the end have trouble getting them out of the water. This is because they will become so intrigued, especially with how it makes their skin look and feel, even smell.

Trains have the sounds of metal and the line up look. Trains are unique as anything and really can make your child with autism so happy. I have trains myself. Going to train places are so exciting, I could spend hours watching the trains they set up all around the rooms. Taking your child with autism to a train park can be so wonderful and many people at these train places are now getting used to having their place be a hot spot for autistic people. They enjoy it when they see your child or person you are taking care of with autism light up! Many of the staff who work there actively include them in their interaction with the visiting public.

Tops are great favorites of children along with those things that spin beautifully and go in different directions in the air. Things that hang from a fan are so fun to watch. I spent about thirty minutes spinning them outside a little shop until my parents came to take me somewhere else. I would have trips with band and I would be staring at water fountains and those fans. My great teacher would always be there to keep me on track, as some of the teachers participated in the college where I then went and LOL, took it upon themselves to take charge again! There are also these things where you drop a penny into a funnel type thing at zoos, malls, etc. that are also a great fun for people with autism. Spinning in a swing can actually make the person feel better and does not result in dizziness when confusingly, walking in effect can. Strange, but, "Hey, if it works and feels good, why not?"

Fidgets are wonderful and they can be small to carry around in your purse to help ease a meltdown, prevent one, ease their anxiety, produce concentration and calmness, etc. Some of the things I mentioned above like tops could be a fidget as well as small mazes, your water therapy CD on a music playing device, straw, etc. Go to my website and visit the corresponding page about fidgets, which would be on the adaptive sensory page. I have a list of websites and ideas for you for wonderful fidgets!

Before going onto the next chapter, I want to let you know that many people with autism who crave input will swing their arms a lot around and about. They can do this going from sometimes to almost all the time. (*This can be screeching the voice too, which I'll bring up after this*) This can be ok if it is not disrupting a classroom, etc. but sometimes it can be disruptful. For that reason, I will give you some tips on how to help with this craving.

I would suggest that you get one of those workout type balls that don't cost too much that you quickly fill up with air (*the plastic air pump comes with*). Then, you can have the autistic person with this arm waving craving wrap their arms around the ball and sort of lay on it. This can help maximize the input they need because of how large the circumference is when he/she stretches their arms around it. This could help decrease the need to swing them about at school during those times that would be very distracting to others. In addition, you could and should let he/she do the arm circling after doing the technique with the ball as it can finally relax those muscles (*like before school, during therapy time, after and before bedtime*). It depends on how deep the craving is. Without the input, the arms can feel like dead weight hanging there and sometimes tingly almost. If you were to lie on a raft in the water and lay motionless with your eyes closed and in complete silence, you could sort of get a sense of that (*and believe me you will at least soon want to swim with your arms to get feeling*).

Another thing that could be added to the regimen for this is a z-vibe and tip that vibrates. They have a very soft humming sound, so that is very good because it's not like some of those vibrating pillows or feet massagers that are so loud. He/she could get input through their hands from the vibration of the metal stick and z-vibe tip that will go unconsciously all the way through the joints of the fingers and to the muscles in the arm that need feeling. These things combined could help to lessen it if it is a craving issue, and even if it is not.

I will quickly now bring up what I said I would earlier, which is referred to as high-pitched screeching/shrieks with the autistic person's voice. This screaming in many is mostly because of needing to feel the upper chest. It is also because it is for hearing stimulation that is needed through a compact and controlled area (*the inside of their own body*). It can be like a sneeze as well, as the feeling needs to come out because it is sitting there in the throat. It is waiting to be stimulated or stirred up through a controlled compactness (*front and back of throat touching each other*), and the special nerves that go through that area of the throat; the glossopharyngeal nerve exiting from your skull. The reason for this is that, that nerve is so different from the others, for it goes through complex paths other than just nerves under skin. Also, if the person is nonverbal, it is their unique way of communicating. You should try not speaking for a whole day. There was an experiment that a guy did to see if he could truly transition to no vocal sensation through conditioning in a lab type of place. He lasted about a year, but what the people later realized was that he had his vocals compromised so that he couldn't feel the sensation or the wanting to use them. It later was reconstructed or regenerated. I really hope this can help you to understand. It is sort of a sympathy technique, similar to what one would use for people who are living with a person who has gone blind. A sensory diet and allowance of screeching to happen when absolutely possible can help to minimize it. These things do not stop cold turkey. You can use some hot sauce to help which is what works the best in polls and studies. (This is not used as a punishment but as positive stimulation that could end the screeching)

I truly hope that you will look into these techniques as a way to calm and help even fix some difficulties due to the autism!

15

AUTISM, THE VESTIBULAR SYSTEM AND ADAPTIVE SENSORY

To start off, many people do not quite understand the word "vestibular." The word "vestibular" refers to balance and equilibrium, which in turn affects other sensory issues as well.

Vestibular therapy is so crucial to the brain connections in people who have autism and similar syndromes.

There are usually three areas: the inner ear, the tactile stimulation and deep pressure.

The body needs to move!

The Inner ear: When an autistic person's vestibular is off, all of their sensory is off and it feels like an "Alice and Wonderland" type of feel. You can feel tippy and can be dizzy, along with things feeling strange. You cannot tell where your body is in space so you feel floaty and in turn can have meltdowns. In addition, it can feel like your body is on a freefall ride or on a roller coaster, but your body is still. It can feel like your head is not attached to your body. It is I think the most uncomfortable and scary feeling. Sometimes, holding the head of the autistic individual with your hands during a meltdown will help with the freefall feeling and help stop feelings of imbalance. This is also where muffed headphones come in. They help with noise reduction and the pressure on the side of the ears and head for the vestibular system issue. My dad has headphones that are used for hunting and shooting. I have used them sometimes when it is like, "Ok, I need noise and stimulus reduction right now!" In addition, miracle and weighted belts (*as well as pressure belts*) can help with this as well for grounding. You can find all these products, where to buy them, and videos of them on my website. Many people with autism will have meltdowns in cars because it feels like they aren't really hooked in a seat belt. It is also the feeling of the jerking and stopping of the car. I tell parents whose children are having meltdowns in the car to try and see if they or another person in the car can find the trigger of their child's meltdown.

Meltdowns in cars are most often vestibular related. Flapping of the hands in front of the face or an oral vibrator that can be held onto can help when in the car.

For example:

1. Is there more melting down/nervous when the car is going faster?
2. Is there more melting down/nervous when the car is going slower?
3. Is there more melting down/nervous when the traffic is inconsistent? Meaning – Your speeds aren't flowing naturally and you may have to keep breaking or pushing down the pedal hard, then speeding up, etc.

Some children will find that certain speeds can help and fidgets (*however of course you can't control your speed to their level all the time*). Also, sometimes it is the sound of the air brushing up against your car windows and underneath the tires (*shocks*) that they can hear, which in turn affects their audio sensitivities a lot. Having cotton in the ears can help with both the speed correlation and those noises. Also, them seeing the sun reflecting in the windows can be painful for their larger pupil eyes, so small car window shades (*like you see for babies with designs, etc.*) can be put on. IN ADDITION, it may help to see what part of the car upsets them the most, like the front or what seat in the back. They could feel too much of the rush in the front seat or they may feel too much of the tire bumping on the road in certain back parts of the motor vehicle. Sometimes the bumpiness can feel good, and vice versa. You certainly need someone along with you if possible to help because you cannot really document this all when you are supposed to be concentrating on the road. Get someone to help you view, or rather them, view these possibilities and test them out! I know it sounds like a lot, but observation is key in autism and figuring out what can help.

An incredible site located near Minnesota that provides unbeatable costs for therapy swings for in-home/doctors' offices/schools, etc. is awesome. They travel to several of the autism conferences and have even better package deals during those times. Check it out. It is (*check my website under adaptive sensory*). Also, check out the same page on my website for pictures of these swings and mounts in our house as well as videos.

A mother had a child who was acting up on a boat, which resulted in things being taken away and leaving early. She wondered why he didn't seem to listen to her at all. I explained these things to her:

All boats have a movement, even when not being ridden. In autistic people, this can cause "wrong" connections and can result in throwing off the vestibular system (*balance and coordination, even coordination of the thought process*). So when the parent said that the child was not hearing her, it was that the movement was interfering with his ability to process verbal commands properly because of the vestibular being so "off." It is not that he wasn't listening, but his thinking process was not working. So, the child could hear what the mother was saying, but his body was so thrown off by the vestibular system in autism and movement (*of course as well as stimulation*) that every word coming in was maybe becoming "twisted" or "confused." Or else, his system was so "off" that none of the spoken words were going through the cause/effect process. All that mattered to his body and brain at that point was balance, so everything else kind of in a way shutdown temporarily. Also, it is actually a good idea to not take things away (*like you would with a typical child*) but actually give fidgets to them that help to de-stimulate their bodies and balance the brain and vestibular connections. Then, he or she can become more stable and be able to process things more correctly (*instead of it seeming like a foreign language*). Whenever you deal with autism, almost everything is different in terms of dealing with situations.

Also, sometimes the child will be able to handle another boat (*diff. size*) as well as one that is going into a lake, because the current underneath a big river or ocean is of course much stronger than in many lakes. He/she can stop when they want as well and just hop out to get it out of their system before it gets to be too much as it was for the other boat. Hope this helps people understand this! Now to Tactile.

Tactile stimulation: Sensory from the eyes, ears, mouth and stress is connected and can cause those feelings above as well. It is a very complicated, confusing thing most of the time. When dealing with vestibular and sensory issues you will incorporate the use of fidgets. You can get real wonderful tactile stimulation fidgets and equipment, along with learning toys at (*check my website under adaptive sensory*).

Fidgets for tactile stimulation are great and are wonderful stims that, depending on what they are, appear to be "normal" in the typical world. It is more of a not-as-noticeable type of stim method! There are of course fidgets that should not be thought of as abnormal, but I guess could create a stare. However, that should not matter. Take note that lots of kids always bring stuff with them, even big things like dolls and strollers! The three sites that I had mentioned under the tactile stimulation paragraph from my site, I have personally ordered from for myself! Many adults love the (*check my website at the same page I mentioned above*) site as well! They order these things for office parties to make their desks more inviting or use the fidgets and objects when work is moving slow.

Deep pressure: One great device, called the squeeze machine that was created by renowned Temple Grandin, is awesome for this. Also, wrapping one's self up in a shower curtain (*I have always done this*) is a great way of input. Joint compressions are another. (*More about this in Autism Therapies chapter*) Also, be sure and refer to my website to see videos of the squeeze machine in action. When I was younger, I used to ask my brothers to lie on the cushions that were on top of me and I used to try to crawl under the couch. At school, in the lunch pits, I would lean over the railing a bit to get that sensation. I would also stretch my body out between smaller spaces to get input through my heels. I would actually bang them into the wall. I of course did not know why I was doing that until I learned about it. Some Occupational Therapists as well as autism advocates have actually been learning from me! They are grateful that I am verbal and are connecting me, what I am describing and how I am reacting to their other patients and surveys, as well as workshops. Only a few dedicated therapy places use this machine, as it costs a lot of money of course. I wish it could be utilized more through insurance and fundraising for more therapy places. There are devices similar to this, in which the child or adult is underneath rolling bars (*safely*) and it gives sort of the same effect. It can be bought on therapy product websites for all disabilities and of course autism.

I want to mention that to help get control of the trunk and other vestibular centers, to make the child almost want to work, you can have a game that is interactive that you find they like. Therefore, they can do that game and it can encourage them to work harder because they see that when they do, they are getting to play the game. It depends, so try to find things like television games that are interactive or simply make up your own. TV games tend to have a very addicting nature to them and can be very appealing to the autistic person! You have to make sure beforehand that there is not too much strain and that a seizure disorder will not emerge from the TV game.

I hope you take what I have given you and use it to your advantage!

16

AUTISM THERAPIES

Since the beginning of autism, people have found natural ways of helping the symptoms until dedicated scientists started utilizing what was natural to the autistic person and turning it into a therapy.

I first want to say that sometimes finding certain therapists is no picnic to say the least and is very frustrating. I do have a few suggestions:

1. If your child is in public school (*meaning preschool through senior high*), go through the psychologist there.
2. If you are an adult it can be much more tricky. One thing you can do is that if you go to college, talk to the DRS there. Explain your situation, and see if they can't help you get in touch with the people you need to get in touch with.
3. Go to the back of your insurance cards and get the number they have and call them. Speak to a real live person and explain your situation. They have a directory so they should be able to help.
4. Sometimes trying to find a specific therapist at one time is too hard and too difficult because of just that one therapist's strengths/weaknesses, experience, etc. So, an alternative that can be much quicker and more successful is going to just a

normal psychologist the first time. This is because they should and probably will have all the correct resources automatically. In addition, when the psychologist is going through the list of the specific therapists you need and then recommends one, they typically know the referring therapist personally. If not, they usually at least know the success of their clients with them.

I have listed the therapies below and will explain each one in detail as they are in a row: (*I have an extra one at the end*)

Occupational Therapy

a. Swings – Used to work on visual tracking and improving gravitational stability (*knowing where your body is in space*)

b. Surgical brushes – Used to help the nerves under the skin to desensitize. You brush on all sides of the arms and legs, 10X each side. You brush down to calm down the system and up to "wake up" the body. I have been using it for quite a while now. I notice that when I do not do it sometimes, it is like a drop in the system and going back to it helps that. It does not work for everyone, but it is working for a ton of people. (*Do not brush the feet or stomach*)

c. Oral vibrators - Used to encourage jaw and tongue movements and help to increase a child's food selection. In addition, these things help the autistic person's throat to be able to feel when food is going to slip down the back. That would help to prevent choking. I like to describe this as like tactile "suddenness."

   It also can also be used to hang onto when upset for needed input through the hands, etc. It works great for some and very much works for me!

d. Weighted vests – The use of the weighted vest provides the person with autism unconscious information from the muscles and joints. It works really extremely well with people who tend to become very easily distracted, hyperactive and cannot concentrate. It gives them proprioceptive input and stability. I used to love to wear a vest, as I felt grounded, secure and much more regular. I still do. (*You can also have weighted belt! I have as well*)

ABA Therapy

a. Reward system – Do something right (*even small steps*) and give lots of praise and stickers, etc.

b. The three-step system – I will give you an example below of a situation in which you would use the ABA three-step system.

Sometimes ABA therapy is overdone...

Here are some simpler ways to look at it below.

ABA therapy:

1. What is my child doing that I can reinforce? EX: Maybe he takes out a fidget when he is getting nervous instead of having a meltdown.
2. What do we need to work on or eliminate? EX: The anxiety needs to go down. Maybe we are moving too quickly in trying to promote speech.
3. Can we provide alternatives? EX: Maybe we just have not thought about letting him swing for five minutes between each therapy and teaching session. Maybe we can offer him a weighted vest to take some of those damaging nervous habits away.

Ask the therapists and teachers about this type of ABA in their mind. We used to do this in our ESAP class in high school. I know that you are most likely wondering what ESAP stands for, so I will tell you. ESAP stands for Emotional, Social, Adjustment Program.

The psychologist would help us come up with ideas, such as having a small picture of something we really like such as our pets. We can take that out when we are getting sad or angry instead of having a meltdown or reacting negatively. It has helped, as I keep a picture of all my animals with me! We called it our "ABA toolbox."

Dietary Supplement

a. Gluten Free diet **

b. Casein Free diet **

** Gluten and casein can act like opiates in the body. The negative effects can occur in people's bodies and start profound bowel and yeast issues. The thought is that, as the yeast grows uncontrolled in the intestine, it sticks to the intestinal wall making it permeable. This diet does not work for everyone. In fact, some children can get sick on the diet because of the lack of some type of fiber that would fulfill the need if they were eating gluten-filled foods instead of being on a gluten-free diet. This is why you would need to talk to a dietician or gastroenterologist about what supplements to include in your diet to help cover the lack of these specific fibers. The diet would eliminate the gut problems or lessen it at least 80%. In addition, it has been known to greatly enhance speech even in children that were nonverbal, particularly since they would not have to worry about the gut issues that make the body have to work so hard on everything else that it cannot work on speech at the same time. This diet works for some, for others does not do anything, and in some cases does the opposite. I have been working up gradually to see if even some of the diet works. In my case, I only need some of the diet to have less gut problems. However, when I was eating a lot more of the diet I was not feeling well. I have some gluten free and casein free food while the rest of it contains some gluten and casein. For some children it will work halfway and in others, it is either the whole thing or none at all. It really depends on the child or adult. On my site, I have a page dedicated just to gf (gluten free)/cf (casein free) recipes and plans. I would like to mention that when you purchase gluten and casein free foods from www.gfMeals.com, and enter the coupon code "Unlocking" at checkout, 5% is given back to Unlocking Autism, the organization! One thing you can also use to help the gut is a stool softener called DOC-Q-LACE, aka Co lace. The DOC-Q-LACE of course you would want to check with your doctor as to make sure it is safe with everything you or your child are taking, but is fine with most. You do want to drink just a little more water while you are on it. Go to my website and I have an entire page with recipes, details, etc. I have to mention that just recently I have gotten off some of these meds or at least lessened them after using the Jack Le Lane Juicer. It is quite a bit of money but well worth it! You have to acquire the taste, but over crushed ice, can become tolerable and even kind of good. It helps to boost the immune system and you can get in just two medium sized glasses, your doctor healthy body structure recommended fruits and vegetable servings. This juicer comes many times with the recipes. Many of times, you just look up what benefits each fruit or vegetable has and then just attempt various combinations. Sometimes because there are natural sugars in the fruit, you will probably want to have something like some chips or some bread/meat. The pulp can be made into soups or muffins. It looks a little sick but actually is not bad when put into these things correctly.

I want to add here, that since the diet is a sort of yeast detox, I will tell you about another detoxifier. It's aside from chelation therapy (*mentioned later*). Some children with autism are prescribed by nutritionists and DAN doctors things like glutathione cream and certain baths (*like clay baths, and liquid needle, etc.*). I have a page on my website coming soon that will have a link explaining what to expect after the soaks, what side effects can be common (*such as specific flakes protruding from the hair/head afterwards*), tips on timing of the baths, and last but not least, reviews of successes along with stories. One thing I want you to know up front about the glutathione cream (*similar but different to Chelation in many ways*) is that the cysteine-SH group glutathione, binds mercury and also protects essential proteins from functional inactivation. It is a principal mechanism of excess

(*EXCESS*) mercury extraction, which is why that part is the same as the chelation.

Before moving on to another therapy, as long as we are on different oral products as well as diets and creams, I would like to mention B-12. B-12 is common as well in the autism community. It helps with the nervous system and promotes energy so it can help with draining of energy before bedtime (*making sleeping more regular*). It can help with the stimming and tics, other motor movements, and to settle the nervous system that is too hyperactive. In some cases, the nervous system is low, and in a situation like that, the B-12 would help to gain more function of those reactions from your body that are needed.

Joint Compressions – This gives needed input to help the body 'feel' itself. Make sure your professionally trained person does this and teaches you how before you would ever try it. In addition, having a child jump up and down on the floor is doing basically the same thing by giving the input through the heels, as you would do compression from the heels during this therapy as well. It is done through the joints of the body, including the fingers. I have had this done. I do not always like it for I have to be touched, but a therapist can make their own hands freezing cold and then I will tolerate it. It really does its job after only a short time. I just feel whole again, so I do not have to feel like I have to push myself into a wall or mound myself to the ground. There is a video of this being done on a child on my site, so check it out! Sometimes they are called "squishies."

Auditory Training – Listening to processed music for ten hours is an intervention to help with speech and sound sensitivities. One auditory processing difficulty that occurs is when a person hears speech or sound, but does not perceive the meaning of the words or sounds right away. I may try this, as I have slow brain processing, similar to the way it was phrased above. Also, this is why I hear what someone is saying and then about 30 seconds or a minute later, I'm still processing auditorily what he or she said 30 seconds or a minute ago. Therefore, I'm missing everything they are saying at the present time. A picture machine or writing machine that shows everything the person is saying is visual so you do not get behind as much. However, these are used in schools and cost a fortune. I used to have a teacher who would get extremely angry when we were reading aloud and I would get lost. In addition, she would get furious when she would ask questions to students about what we read as a class together, because when she would get to me, she would always accuse me of not paying attention. She said that if I did not care then I could just leave. I could not comprehend at the speed they were reading. I told her this alone and her response aloud was, "What? I am reading at a special education pace even for kids who are Dyslexic. If I talked any slower I'd talk like a robot." (*She was my special education teacher for goodness sakes and everyone in that class was nice to me and always helped me out, even going to the people at the school themselves I guess to let them know what this teacher was like. However, she is still employed and upsetting other students I know*)

Listening Therapy – This is a therapy for tolerance and awareness of sounds, touch, vestibular/balance, etc. When this is used, especially for vestibular and balance, the listener must be in a more settled down area without a ton of light. They can be doing an activity like brushing the dog (*but cannot read or watch television*) while listening through headphones for 30 minutes a day/ two times a day. You go through a series of specific CDs by an occupational therapist because these CDs are made with vibrations, sound fluctuations and brain/ inner ear trainers. Do not be at all surprised if you have to keep restarting the process over at least in the beginning, because I would keep taking off the headphones because it was too much for me. In addition, it can make the sensory a whole lot worse before it gets better since it is almost like going cold turkey. It would not be a problem for a neurotypical but for the autistic brain, it is. It felt like a bee sting anytime anyone would touch me and my sensory was pushing me into panic attacks for a while, which is why I was modified and had to keep starting over. Sometimes, the starting over is not terrible as the body is trying to adjust temporarily before going completely cold turkey on its own. However, my therapist says it is worth it and it really will help. I do not know about that, but I guess I will just trust him, maybe! For me, the system that I am using prescribed by my occupational therapist is called "Vital Sounds." You need to go to your therapist and they will write down how you are supposed to do it and what CDs in what order. This is very crucial that you follow it exactly.

Support group - These groups are great, as you form a network of others who are just like you. Therefore, you accept one another and help one another by sharing ideas, etc. Before I got sick with Valley fever, strep throat and mononucleosis, I was going to the support group. It was great! It was sort of like my family that I had at the high school. I went back to the ESAP room a couple of times just to join again, especially since some of the people were still there who were not yet graduated. It is awesome. A few of the people in the group used to tell me for some reason, "You are more autistic than us." I think what they meant to say was, "You are more affected than us in certain areas." However, we all had some part on the spectrum so could relate in some ways and supported each other. I am happy to say I am now getting back to them a little. They are good for helping each other out and there are also meetings such as bridging the gap (*helping people with autism and their families understand each other better – sometimes split into two groups*) that can be funded through your own Autism Chapter or community. These can have older but not that much older adults on the spectrum who can share with the younger adults what has helped them in life. The younger adults can also teach the older adults new things to maybe try. It is just so neat!

Hyperbaric Oxygen Therapy – This is used to help feed the brain extra nutrients and to help pressurize the body correctly. I will have a video of this being utilized with an autistic child soon, so check my website for that update!

Chelation therapy - Believed to help remove excess metals out of the body. (*Please check my website for videos*)

Neurostimulation therapy - New! This is used to redirect affected parts of the brain (*such as sensory/brainwaves*) through metal circles they stick on the head. They can then make a therapeutic CD to help with the brain waves associated with autism, slow processing, visual distortions and sleep issues. Check my website for the video of an autistic child getting it done and parents talking about the benefits! Also, see the autism and sleep chapter for a very similar and very detailed description of this therapy.

Horseback riding therapy – This is awesome for the vestibular system, muscle tone, silent connections, etc. Check my website for a video of a child with autism utilizing this therapy and hear the benefits from the person who works at the therapy place herself. This therapy is many times called Hippotherapy, partly because it can then be considered a real therapy medically as well. It can have an actual prescription written out for it too. When your doctors write for example, "AMOX CLAV" they would be putting, "Hippotherapy" instead. For me, when I am on a horse, I actually start laughing because the bouncing from the horse feels funny.

Chiropractor therapy – Going to the doctor's and getting a referral for a chiropractor for autism is wonderful because chiropractors can work with the nerves (*which are a major factor in autism*) and (*again not so commonly talked about*) pelvic displacement, or pelvic slipping. You hear many times in a dx guideline that the person with autism has a "strange" gait or walk. The reason for this quite many times is because of the pelvic placement in some autistic people, as it can slip even millimeters out of place causing the walk to progress slower. It also can cause a clicking of the bones and muscles when walking, because it's connected to that area of the body. It is caused by pelvic instability from studies in autistic people and how they walk. This is also connected to the basal ganglia in the brain, similar to people who have Parkinson disease or other movement disorders. The chiropractors can adjust this, and I personally have had this done and hope to go to a chiropractor for the neurology part of it too. You should find a chiropractor that will work well with your autistic child as the person with autism can become easily afraid of that type of stuff.

I also want to mention here that for deep pressure input as well, you can do fun activities for social interaction with your child by having a blanket they like (*wrap it around them while you hold the sides of it*) and then pull them towards you. Many kids will love this and get the craving very well! Also, many will do the resistance before you pull and during, which gives greater input. Finally, release them so they can fly into your arms!

So how do I fit them in correctly without getting overwhelmed myself and overdoing it for my child/sibling with autism?

1. The first day of O.T. and some more after will probably be overwhelming. If you can spread it out during happy times at home as well, that would be good. Make sure they are taking it slow as doing it too quickly can turn it into a bad experience and make the autistic person afraid. In addition, it can help to have the O.T. make their hands cold instead of a regular at normal room temperature hand. It of course will depend for the individual.

2. Rushing is not good... overworking is bad too. In school, it is good and spread out throughout the day. In the school or outside the school your therapists should be doing brushing, ball therapy, lotion therapy and swinging therapy but in moderation. In addition, you should be in there so you can incorporate the therapies throughout the day, short times per time, but not too much. You would just want to be there a couple of times so that your child will respond more to the therapists instead of focusing on you. Joint compressions are also what are supposed to happen. Visit the school and see what they are starting. If in school it is fulfilled pretty well there, just a little at home should be sufficient. You as the caregiver or adult do not want to over occupational therapize, speech therapize, or over therapize any of the other important

therapies. This is very important to make sure you are utilizing correct moderation when doing the therapy.

3. I would say the constant appointments might be able to be replaced by spreading it out throughout the day in your environment, incorporating it into fun.

Examples:

A. Speech therapy during playtime can be great. Also, you can use the z-vibe and tips before every meal that helps him/her feel his muscles in his mouth as well as his tongue. It also helps to properly regulate many of the gustatory reactions that I mentioned earlier.

B. The brushing can be used as showed to you by an O.T. after breakfast, maybe once throughout the daytime at home, and after dinner. It also can be used before an event and/or after if you know that where you are going could cause over stimulation. The timing affect is measured by the reason you are using the brushing. If the person is using it for the nerves, it can last up to two to four hours. If using it for overstimulation and or calming/waking up the body, it may need to be completed somewhat more frequently if necessary.

C. If you get swings in a room, he/she can go on those during play and he/she will probably want to start going and using them lots.

These are just a couple examples that really give a lot of input and can be done, as you can see, throughout the day; not just within appointments. Therefore, it can be a greater benefit as well when you can take these appointments, really learn from them, use the techniques in the therapy session and "scattergorize" them throughout the day as part of a routine.

I know this seems like a lot to take in but worrying is not going to help. Just take/do these things slow and in moderation. It will work out to all of your people's advantages.

Examples of therapies:

1. Making the body move
2. Brushing
3. Z-vibes and tips
4. Visual interaction
5. Signing for the hands, speech, etc.
6. Joint compressions
7. Rolling an exercise ball across the body
8. Moving the ball all directions while child sits on it (*do not want to start with that first and want to go slow with that*)
9. Listening therapy (*Don't be surprised if you see you have to restart this several times*) The therapist should use time management with this listening therapy technique as well as volume control, etc. so that the therapy can be tolerated and of course work.

I said at the beginning of the therapies list that I would add something at the end. I would like to add that there are medications out there that do help people with autism to come "out of their shell" more and become more integrated with their environment. I am on medications and I wanted to let you know that the doctors said they have seen improvements. I've tried many medicines before these ones that either didn't work or I had terrible reactions to, but now I know there are some that really are helpful. It just depends on each person's system. Some parents are surprised when the doctor asks to put their child on an anti-fungal medication. I am not at all surprised because our bodies do not handle things all that well with the leaky gut. I was on Fluconazole for Valley fever and on the back of the box, saw that it said it can be used for autism spectrum disorders. This should be another sign to doctors that our systems are on the fritz. Then, they should check what it is in these antifungal medications that are making a difference in our blood, urine, titers, etc.

I hope that you will take these suggestions and use them. They help make the hard issues of autism a lot better. Once again, check my website for slideshows of these therapies that I have put together, etc. for the colors and environment around the pictures. These tools will help give you that digital extra to incorporate while or after you read what I wrote about them. It may help de-stress your mind being able to see it in a picture as well. See, looking at pictures doesn't just help autistic people, but also neurotypicals. ☺

17

Animal Assisted Therapy and Autism Service Dogs

To start off, I have several videos on my website of just how wonderful, miraculous and "healing" these dogs are to all people on all levels of the autism spectrum, along with organizations.

What these dogs can do:

Just to start, a few things they can do I have listed below. I am going to be getting a golden retriever because we do not like separating our two shelties. However, we will still take Sasha out occasionally so she does not get so upset about having another dog going out instead of her all the time.

She helps me by starting up conversations, alerting to others that I have a disability, another can help me remember to take my medications, and labs or retrievers can do a standstill if I freak out from something that would cause me to run out into the street. I cannot go anywhere as she is tethered to me. Right now, it is that I realize I have her and she distracts me. She also gives me deep pressure input by laying on my legs and that helps when I am trying to sleep too. She licks me and gives me soft scratches with her little nails for input when I am overstimulated. She is also there so I can better balance sometimes on the stairs when my vestibular is off. These are just to name a few tasks. She is my best friend and the one person I can trust completely. I love her so much and that is why I take care of myself. I used to think about wondering why I was living until we got her. I have grown into some agoraphobia, especially since being sick. If there are many people around, she could walk around as to let them know to back off or rather spread out further from my vicinity.

You can actually train your own dog. I have a link on my site and it has the ways to train your own dog. Then you get the vest from (*Check my website*) as permitted by the ADA, or Americans with Disabilities Act. We had trained Sasha, the one that I have right now, and she is awesome. Our only reason for getting a retriever right now is about not separating the two shelties and the standstill command to help me better. Also, we will be getting it from an organization since I am in rehabilitation right now and cannot go out with it a lot to train it from the beginning. I also need one now that is already trained so it will not feel the separation from our two shelties when it is going out to assist me. In addition, some organizations do have the voucher (*to help pay for cost of dog*) through Assistance Dogs International, but some other vouchers available I tried for earlier were denied on the terms of it being a "psychiatric" dog. Check my website out for more details. The dog does have to wear the "service dog in training" vest for longer before it can wear the "service dog" vest. Also, start similar to the beginning by spending lots of time at places like Petsmart and Petco. The trainers there love to help, especially when they know you are training your dog for service. It is a really great thing. All the places I've been to have had trainers or Petsmart workers come over and try to help me (*and they didn't expect a fee*). I also had other dog owners and other people with service dogs coming over to me as it melted their hearts and they wanted to help as well. You can pay little for the classes and go to Petsmart or another place many times. Many people at these pet stores are becoming aware of the need and the high cost of getting a dog from an organization, so they usually get excited when they see someone training a dog for service work. You also need to get them used to NO table food… ever. Dogs can have certain snacks like carrots or peanut butter, but you cannot have them eating it while you are eating. They especially cannot take it out of your hand during that time, as they will connect that with taking food from others. Taking them out in dog places while they are being trained there is wonderful. Take them all the way through advanced training offered at Petsmart or other places and get the vest started on them right away with a wraparound leash. It is really great autism therapy as well. As well as great therapy, it teaches cause and effect to people on the spectrum who often have great difficulty understanding it through other situations. It also helps increase conversational skills and when the dogs are trained, there are a lot of "cues" that are in sign language and strictly to the point. A lot of people on the spectrum are more like Mother Nature in that way. Sometimes our learning has to start out as "just so" in the beginning for evaluation and catching up purposes. If a mother dog wants her pup to lie down or get off of something, they do not go into explanations but just take their pup off whatever it is they want it off. Then, they make their pup lay down. Those are just a couple examples. The logic or the longer explanations can be gradually taught later, especially if a person on the spectrum is nonverbal at the time or under a lot of stress.

I have had meltdowns that have lasted over 1 ½ hours and once they let Sasha in, I snapped out of it within a minute or so. It has even gotten me out of a bad panic attack within seconds. It is so amazing the healing and connection between people with autism and these animals.

If you decide to train your own dog under the ADA, it is not exactly going to be easy, and you do have to make sure the dog is up to it and is willing to do the service well. It is so worth it, more than words can tell you. Watch the videos on my website and you will be in tears.

Go doggy!

This below is a picture of my service dog (*that we together trained under the ADA*) in her beginner's vest!

Isn't she beautiful? She is such a sweet heart. When I was in the wheelchair for a little while during the really bad Valley fever, she always wanted to pull me! She naturally is a herding dog so she always wants to do a job. Service dogs need to play like normal dogs and can bark and behave like crazy, but when their vest goes on it is as if they know; it is time to work and be professional. ☺ I am also wanting to add that she got another patch that says "Autism service dog," and there are "Autism K-9 patches," etc. We got ours from a place called Pup'parel. Get it? Apparel, for pups! I just think that is adorable how they came up with that! They have mostly stuff for service dogs of course, as that is where we got Sasha's patch. When I get my new dog, they will simply unsew carefully the patch and put it on the organization's vest.

On this page, before I go into more details for you, I would like you to know that service dogs are supposed to get a discount, usually 20-50% off their vet bills because they are servicing a person.

Some doctors can actually write up a prescription to be passed through insurance to get your dog almost free if not completely free, as it is an assistance needed due to medical reasons. It can be a hassle or more of a big hurdle, but if you find the right person, that would be great as well.

I know that earlier I explained just some of what Sasha does for me, but I want to describe them more in detail, as in the words and signs you would use with your dog. (*In order to train for some of the things I explained earlier)*

I am going to give you a couple general statements and then go into specifics and details. A couple other things are the dog jumping at or getting my phobia away from me, and then coming back to calm me down. This also helps the public to understand better what is going on. There is also more independence when crossing any type of street, because if I did not have her, I could be standing and waiting to cross for literally forever even if there weren't any cars. I will now go into details and specifics below.

For the first one, as you are working with a dog for a person with autism, a lot of people on the spectrum need input that they don't always get because they can't have a therapy room traveling along. However, the dog can help immensely. One of the inputs you can get from a dog is the lap-up command. The dog will put its head, paws and some of the upper body onto the lap or the shoulders. This can give some deep pressure inputs, especially depending on how much the dog weighs. The other way to get this input while out and about is for the dog to lay its head or part of its lower extremities on the autistic person's feet while they are seated. This helps balance and perception through the feet.

Another command can be given (*especially if a larger dog – this really has to be done with a higher weight*) to do a standstill in case the person with autism would try to run away from you for whatever reason. You would say "standstill" and then the dog would automatically stand or sit there and not move. This lets the person with autism feel that they are being tugged backwards when trying to go forwards, which alerts them back to before they decided to lunge away. It also gives them input that can help relax them. It is sometimes a real safety issue and for others it helps them to not get lost, etc. (*I can get easily lost – another good thing is the dogs have good sense of smell and understanding so they can help "un-lose" you*) The belt needed for this is a wraparound belt. The belt hooks to the dog (*you will want the dog to have a harness so when it does a standstill it is not on the collar where it would choke them*) and then it is like a regular leash except for the fact that it wraps around the waist and has a clip. They are usually adjustable. Sometimes you may truly need to have a harness or vest for the leash on the person if they are really strong and the clip might somehow come undone. It is like a pulley 4-way affect, so it helps with the balance in traction for the clip.

This brings me to another command that can be given. Many places teach it with a snap of the finger and the word "go" with your finger pointing to the object or child who is going away. The dog is basically taught to prance slowly or sometimes have to run to catch up to the person who is running away. The dog circles them or blocks them and sometimes even pulls a belt, part of the great wraparound leash, etc. to pull the person back to where they are supposed to be.

Another command great for these dogs is licking if the person with autism becomes alright with it. Some citizens with autism might not like that feeling, but for lots they become used to it and even really like and appreciate it. The command is the keyword "lick." You can teach the dog where to lick, such as using the keywords "lick hand" or "lick cheek," depending on the part of the body that helps calm them down the most or redirects them. It depends.

You must work with your service dog on sitting, down, wait, come, stay and all of those other commands every day, especially if doing this yourself under the ADA. There are of course the other commands I listed above that are equally as important. You should start taking your dog out pretty early in the beginning but not too early. You need to see how your dog will do and the service dog in training vest is needed for longer than when you get one from an organization. A great place to start as I stated earlier is Petsmart or Petco, and many times, you will make great friends there as you all start to know each other rather quickly. I would not take them to a restaurant too early, but after you've mastered some well commands and especially the process of the "down and stay" command, it is fine. You need to test this out like organizations do. You should really try this at home by having the person with autism eating (*where the food is within the dogs reach*) and make sure the dog either sits if they want to some of the time or lays down. They should not look at the food - rather ignore it. However, some may look at the food but not be allowed to do any more. It is part of the "leave it" command. You can give them their special doggy treat. I just recommend the Charlie Bear low calorie treats so they know that they did what was asked of them. However, lots of praise should be given more than treats after a while because they need to be willing to do it just because and not expect a treat every time. You should make sure they are potty trained and you can also teach them to nudge a water canister or sometimes they have this in a pouch and will nudge that in order for you to know when they need their water. Just ask a person/trainer at a pet store what kind of water canister would be best for your situation (*service dog*) and even ask some questions or advice. Just offer them some water when you have reached your destination, sometimes in the middle and when leaving the destination. I want to mention here that if you live in a hotter state, there are little fans that you can hook to something so that your dog gets A/C. You need to keep them cool. If you have a van, the great thing is they have it like in airplanes so there are the A/C fanners like in the front for the pilot. Plus, as an extra, they have a big nice space! The dog can also be taught to give a command when needing to use the restroom. Some of the time, you may have to ask the place where your dog can go excrete. Many people have "potty bags" in the dog's vest pocket to clean up after them sometimes. Make sure your dog has gone to the bathroom before you go somewhere and most dogs are able to hold their bodily symptoms for quite a while as well as go on command. Usually a tree outside places is part of nature, so your dog can eliminate (*pee*) there and it will soak into it. Besides, there are workers who tend to those areas to keep them up. In all fairness, if they do have a bowel movement, you should just pick it up with the doggy potty bags that are in a round canister and hook to their lease. Some places have an area set up for your service dog that they actually have someone clean up after all the service dogs. I know this may sound like a lot but it really becomes almost automatic and not such a big deal once you have done it a few times. There is always a beginning.

I want to mention that for traveling, if you were to make a flight ticket, you need to tell them up front that you will have a service dog coming along. They are not supposed to charge you extra for that. When you are boarding a plane, the attendants usually give an announcement about anyone with special accommodations (*like baby strollers, oxygen tank, service dogs, etc.*) to come first to board. In addition, many times for more room for the dog, you could automatically be put in 1st class and not be charged the amount for 1st class because it is an accommodation. That depends on the airline and situation.

In my last part of this chapter, I would like to add that you guys should have no access problems in almost any place you go and that is including restaurants. While eating, your service dog usually lies under the table (*if it is big and roomy*) or a lot of times will lie next to the table, just outside of it. On my site, I have a link where you can print out an access card with an 800 number on it with the laws and regulations. It lets the people know what they can ask, what they can't, why they have to let your dog come in and if they do not what will happen to them. Many times, you should not run into too many access problems. Go to my website under the appropriate related topic and I have a link telling you everywhere your service dog is permitted but where there are a couple exceptions. It has things to help you prepare for each destination if you are just learning and it is unique. The places they aren't allowed access to would be the Operating Room and Dialysis. All the other places of the hospital,

ER, etc. they are allowed in. Some hospitals even have these dogs go into trauma units as well to help with animal and humanity-related therapy healing!

Concluding this chapter, I would like you to know that if your dog is able to lie down long enough for a college course, your dog does not have to be approved and listed on your INF forms as an accepted accommodation. However, your instructors in all fairness should be warned in case of allergies, extra room, etc. The college DRS cannot say that you have to have that okayed to be as an accommodation. It is against policy and against the ADA access guidelines. They are there to be with the person so they are to go almost everywhere with that person without prior permission.

I hope that this chapter has helped you understand more what you will encounter going through this process and the wonders it can do. I moreover hope that if you do decide to do this that you have fun doing it. I also hope that you go to pet stores for help and great training, for both dog and person!

I know I'm glad I did.

As a special taste to this chapter, I would like to share with you a poem that I wrote for my wonderful autism service dog. It is below on the next page:

Dear Sasha

I don't know what I did to deserve such a wonderful miracle

I guess it's because we were meant to be together

You do so much for me and I wish I could do more for you

When you stare at me in the eyes you don't see a disability

You see a person who loves you back and someone who needs your help

If you understood all my words I would tell you that you are my angel with 4 paws

I've gone from being able to do things with utter struggle to being able to conquer the world

And I have the best friend and partner in the world to do it with

I thank God for you every day Sasha

I couldn't imagine life without you

For I really didn't have life before you and now I love it because I have someone

And someone who will never be unfaithful but someone who is more than faithful

A friend and a beautiful dog to lean on whenever I need you

Thank you Sasha

I'm so lucky to have you as my Service Dog, as my Friend, as my Angel

18

AUTISM AND PHOBIAS

A note: These ones I am going to list are the main "autistic phobias" and are represented in most movies about autism. Some people are actually calling these phobias, "autistic phobias" now. It is kind of weird because other people who aren't autistic can have a couple of them too. However, it has become so pronounced in millions upon millions of autistic people that they are referring to these phobias below specifically. Moreover, one of them is by almost every Asperger's person.

There is only one more phobia that I am adding on my website and that will cover the five phobias that are pronounced in autism, as being typical for having autism.

1. Autistic people are so many times afraid of butterflies or other flying things. Some autistic people are fine with butterflies/ other flying creatures. The reason is that their wings, when they flap, have a very horrible sound as well as they flap too fast causing overstimulation. If I were near a butterfly, most people would think I am severely autistic. Things will be in a wreck because I flip so terribly. Things will be knocked off shelves accidentally if it gets into a confined place. I was at Golf Land a few years back and there happened to be a butterfly. I started screaming (*sounded like a hawk*) and I actually threw myself to the ground. It obviously caused a scene and I was

not like a little kid, so I'm sure others were really wondering what was wrong with me. At the age I was, most people would not throw themselves to the ground and scream so loud and so strange. There was another instance at the grocery store where a butterfly had happened to get in. I was freaking out, ducking, and of course started stimming. These are times that it is great to have a service dog so that police or security will be alerted and not have to wonder so much as to why it is happening. In school, it was maybe around 3rd grade, they had a butterfly project. There were butterflies in a bucket with air holes and they would have us all go outside the class to let them fly away. This lady was there and I walked the other way before they were released because I was scared and it would be a complete disaster if I happened to be present at the actual releasing time.

2. Not mentioned as much, but very much a phobia in various autistic people is certain letters of the English alphabet (*sound particularly*), numbers, and certain punctuation marks in the English language. Some kids will be constantly getting their arithmetic problems wrong and get punctuation points taken away, but what many people will not think about is "why." As far as the refusing to write a number or an authentic punctuation symbol, some autistic people get upset over a color. In this case, it has more to do with the brightness or what they think of it as referring to a texture. This issue would be caused by something else that was that color and had a negative effect on them. Autistic people are conditioned very easily. Just think of the conditioning experiment with little Albert and the white mouse. Many, many teachers have emailed me saying, "Oh my [gosh], I really thought it was just my student's funny strange little quark or stubbornness!" Also, another teacher said to me, "Wow, I always wondered why my student used to get so angry and go up and erase the dots above my i's or at the end of sentences and replace them with a star or heart. I have noticed more over the years that several upon several more of my students with Autism/

Asperger's/PDD have done that, but I never really connected the reason behind each of the cases until now. Thank you." I want to let you all know this was a major problem for me (*all the scenarios I stated above*). I would actually cover my ears during certain sounds when the teachers would be talking, or other kids, and actually got very angry with my parents when they were saying specific letter sounds. I was not able to tell them at the time why until literally a couple months ago, as this book has helped me to get it out. I can have my parents read this or I can read what I have typed and sort of script it out. I didn't have the number problem but did have the punctuation problems as well. The teachers at first thought it was a girl thing (*because I would replace those specific marks with cute drawings*) but then it became more evident. I actually started to have problems filling in tests (*the bubble tests*) and that is where it was evident that there was something beyond it. I used to get points taken off sometimes in high school because I would refuse to sometimes even say the word with the letter sound (*like for responding on an answer for literal sense synopsis*). I would sometimes mumble over the word in the reading so the issue was not clear. It was so bad that I would refuse to punctuate my papers in the correct way. These are real, sound strange, but that is the way the brain with autism can work a lot. Your child may be melting down and getting aggressive. I hope that these phobias, if they are having problems with them, can alert you to another idea of why they may actually be getting so aggressive.

Other parents have also recognized that their child will get aggressive or angry when they are talking and only started to realize now that it is the sound of certain letters in words that sound negative to the brain of their child with autism. In addition, it can also be not just individual letters but some words just don't mix right with the brain. This is yet another neurological struggle and hard to explain. For each child it can be different of course, but many of the letters that if the autistic child has a problem within this phobia area, are the letters P, S, B, and V. These are seeming to be the most common letters associated with that.

I also want to let you know that in groups people have come up and said, "I feel like I want to tell you something about myself." (*People who would have autism and Asperger's*) They will be at conferences and express their problem, which in quite a few cases, the problem happened to go right along with these phobias associated with autism. They were relieved to know it was not just them and something to be concerned about.

Not only has this book and my emailing others through my site with hundreds of questions help me to also reiterate that it wasn't a separate problem beyond autism, (*although I knew it was not but always felt so confused*) but shows again that it is directly related to autism and its many various effects on the brain.

3. The size and thickness of things have been known to bother many people on the spectrum. For some it is big and round things like pots. For others, small and long things like railing bars. It is very hard to explain. It is a phobia that affects many autistic people. There is a representation of this in the movie, Silent Fall. Look on my Autism Movie Trailers page on my website and I will be adding that clip soon of the child with autism having a meltdown when being exposed to round food. It was when he was in contact with green peas.

   I have had problems with this. Every time we would go somewhere and there would be something like plant pots by windows or by benches, I would try not to look at them and it actually made me sometimes sick inside. This is not uncommon either. It really is a horrible thing to have phobias, but these types of phobias are literally impossible to get away from. However, because of this, conditioning and as time passes, these things can become easier on the person. I recently had a mother who emailed that her son gets aggressive over things like spaghetti straps on shirts and other skinny things. She also was relieved to know that it was a part of the autism for her child and not something else on top of the autism.

4. Fuzzy things really affect the eyes of autistic people and therefore can bother them so bad it becomes a phobia. However, it is a specific type of fuzziness, but not like fur necessarily. More of an example would be like those fuzzy peaches, etc.
5. Last phobia coming soon to my website, so be sure to check it out.

19

AUTISM AND VISUAL DISTORTIONS

1. Sometimes, things can look abnormal, like one chair leg may look fatter than the rest. It can look stretched or even seem to be missing areas, similar to the effects of bad migraines. Many times a child with autism will be diagnosed as having some problems with brain waves (*a common autism symptom*) and they can be put on medicines to open up more vessels in the head to help this. These disturbances in the brainwaves, although in most cases is not dangerous at all, is inconvenient and of course annoying.

2. A lot of autistic children look like they are grabbing the air and staring at it. The particles in the air that are supposedly not able to be seen except under specific light, are seen by autistic people since our eyes are much more sensitive, like animals. One science guy asked me what I have seen. I explained it perfectly as to what I was seeing, and he was able to tell me about experiences working with lights and its affect on particles in the science department and when I checked myself, it was indeed truly fascinating. Another reassurance was when talking with other people with autism who said, "I see white stuff." That is always how I tried to explain it, although to me I honestly thought they were flies or some creature. In a bit, I explain how distressing it was when I did not understand

that what I was seeing was just part of the atmosphere when I was younger.

3. The next visual distortion in autism is called "tornado" sparkles that happen when looking at the ground and concrete. This is because of the chemicals used in making the sidewalks and streets that creates this bizarre visual especially when friction hits surface. In addition, this is another thing supposedly not able to be seen so easily, but autistic people can.

Before I go on to the next paragraph, I have stared at the ground funny and people did not understand what I meant when I said, "It looks like a tornado sparkle." However, my science teacher knew what I meant after I told him and explained why. His reply to me was almost identical to the explanation that I gave you above in number 3 from a different person. This also goes along with sometimes when your dog is barking and hitting their snout into the ground. They are seeing what a person with autism sees, or in reverse, the autistic person is seeing what the dog or cat is seeing. Many cats will sometimes jump over a spot in one area and this can be almost completely the reason in cases. This is again another huge similarity between autistic people and animals. We are very sensitive in several areas.

Most of these visual distortions really scare autistic children when they have no idea what they are. When I was little, I would cry, hide behind the couch and self-calm because I thought the air particles were going to attack me. But, I have learned to deal with them, as I found out the reason and the scientific explanations. In addition, after 19 plus years, you start to get used to them. However, they create attention problems, as it is hard to forget they are there when you can see them all the time.

All of these have been shown reputable from feedback and review, etc. that these are not just what I just see myself, but what every other autistic person could see too; <u>that is depending on how much autism</u>

is affecting the person sensory wise. This also goes for the other things that I have written in both this book and on my website as well.
I have posted on my website under "What does it feel like to be autistic?", two wonderful videos that everyone here would benefit from. One of them shows a great representation of visual distortion. This also goes for auditory distortion as well, which is represented extremely well. It is also an extremely good representation of auditory processing difficulties.

I hope these have given you insight into what we are seeing and feeling on a daily basis. Also, I hope it helps explain some of the behaviors you see in your children. Many parents have emailed me thanking me because they were terrified their child not only had autism but a psychiatric problem as well.

20

AUTISM AND THEORIES

I have the theories of mercury (*the big debate with numerous directions*). I also have included a smattering of great possibilities including genetically modified foods (*sounds creepy*), processed foods, and then pesticide studies and autism. I am gathering information on autism and pesticide studies, so I will post them soon on my website. Again, check my website for posted studies, pictures and even videos concerning these subjects.

Theory1: Vaccines cause autism.

Theory2: It's actually the thimerosal/mercury preservatives in the vaccines, and not the vaccines themselves.

Theory3: It's the thimerosal/mercury put into the blood stream in babies pre-dispositioned to have negative effects from metals.

Theory4: The timing and/or amount of the vaccines being given in a time period.

I am not necessarily one way or another on the vaccines. However, we had come to find out later about some individual shots that actually contain thimerosal-based serums in the medicine. I have heard and seen people whose children were entirely "typical" and literally, within 48 hours of getting vaccines, their child was full-blown autistic.

A sweet family even received compensation for their autistic child although they say it was because she had a mitochondrial issue; they argue that the Mito disorder either was from birth or caused from the shots. From all the cases, you cannot just look at them and not wonder even slightly no matter how strong your belief.

I also have to add one thing to the vaccine debate. I have learned through statistics that about nine out of ten times, while going through trials, when bad effects or even death occur, it is written off as "an abnormal coincidence" and the general population is not informed about it. Other vaccines had commercials for them. These commercials kept going on for months when the vaccine had caused numerous heart valve problems, paralysis and even death. Why did they keep doing the commercials? Truly, I think it was just for money and to lighten up what was really happening. This is what people who study these scenarios also assume. You can talk with your doctor about spreading out the vaccines and testing for allergies before the injections (*in some serums there are animal secretions like embryonic vesicle fluid and lung cells; just to name a few*). Also, watch out for the premature babies. If your family has a history of allergic reactions to vaccines and or neurological disorders and diseases, talk directly to the doctor even if they do not want to listen. If they don't take the time, then find another doctor who is willing to put questions and monitoring first, although the if you didn't get the shots, obvious diseases could occur and be deadly. They should be using more natural preservatives in these shots so you can be protected and not have damaging chemicals going into the bloodstream. It is such a risk because you do not want to get these terrible diseases (*there are commercials showing how when you buy diapers you can save babies by them just getting a vaccine, and it's true*) but then you see autism, OCD, asthma, ADHD, and other serious conditions linked to these vaccines. I think that personally, one, it is because they are not replaced with natural and normal things to preserve them (*as well as the bulk of them at once*). If it costs more currency for the companies to reproduce vaccines for better safety, then I think that the people making them need to ask for help if they want the country to be safe from bad diseases, while lessening the risk of these other things I listed above; autism, OCD, asthma, ADHD and other serious conditions.

Most recently, scientists have been looking at a new chromosome that is irregular in some autistic people. However, it is only a handful of autistic people. This is a good thing though, because although it may not be a gene causing autism, the scientists can hopefully see what factors (*meaning foods, chemicals, global issues, etc.*) have damaged that chromosome and go on to test those things as possible reasons for autism. That would be wonderful for many people.

Processed foods were not as available back then to buy. In addition, back then we did not see these cases of autism or as many of other developmental or physical disabilities.

Genetically altered foods take up about 80% of the grocery store, even including fruits and vegetables. On my website, I have the specific labels and code numbers that will help you to identify which foods are truly organic. When the parents have been eating all these foods, it goes into their body and changes their gut-fighting agents as well as nutrient absorption. So, it passes from their body all the way through the umbilical cord and into the baby. This gives the baby an unfortunate start to its absorption and gut-fighting agents as well as their body's own ability to fight off excess metals naturally as part of the other theory on vaccines.

Depleting ozone layers is causing elements to enter our earth's atmosphere and affecting sun activity. The bright sun's activity affects our bodies, and the elements affect our foods, what comes into them, out of them and then into the person. This also affects the lungs, which are in connection to other vital organs causing problems in those areas as well. You will have a chance to read about our moon activity in the upcoming chapter.

On site, I have the posted study, information on number codes for true and real organic food and have a video on it.

So then, why can autism occur in twins?

Answer: The autism link in fraternal twins is approx. 10%. However, fraternal twins share about 50% of the same DNA and gut fighting agents. According to the statistics that I state on my website (*check them out*), babies have been receiving far beyond what they should be receiving, so tolerance is looked at. These tolerance issues are passed down in family generations, so it would then not be uncommon to have relatives and past family tree members that have autism in the family.

To wrap up this chapter:

If you take all of the things I said into consideration, read the studies I have posted on my site along with the pictures and videos, including rat testing, then you will see that several things may be causing autism. My theory is that it is exactly that, because they are bonding together, similar to the way an atom does.

21

AUTISM AND WEATHER

As part of my theory on animals and autism, an autistic person is similar to animals in certain ways. This would be one of them. Animals, when the weather is changing, lay down and some start bucking or calling out to the rest of their herd. It is very similar in autistic children for you to see stimming and different behaviors, especially voice stims.

In addition, the barometric pressure affects an autistic person's vestibular and sensory system, causing more automatic responses from their body in order to keep regulated.

I want you to know that I have videos that show similar behaviors on my website that will help enlighten what I am referring to here in this description under autism and weather.

Some people on the autism spectrum have really good temperature regulation, which is another great similarity among some animal species. It is the flexible adaptable survival ability.

You may see your autistic child dressing in what would be considered inappropriate for the weather, or you may in turn have an autistic child or adult who you see cannot regulate their temperature very well. They would end up needing medication. So many times, you will hear a doctor or psychologist say, "Yep, they have that broken internal thermometer!"

Sometimes, some autistic people will want to wear tighter clothes, especially jeans and not shorts because they need the pressure so their muscles do not feel "sloppy and airy." That means it basically feels like it is not there. Sometimes, it is also because they do not like the feeling of their two legs (*the skin*) touching each other, as it feels really horrid.

Some children/adults are actually "fixed" by rainy weather... it depends on the person. Usually the wind really gets an autistic person upset because of the noise and overstimulation to the eyes. I have even literally hidden in the closet, screamed and stimmed when I am home, especially alone, during windy weather. My service dog helps me there now though I must say I take her into the closet with me at times!

Cold weather that has to involve layering and seams in clothes, along with tags, are really upsetting to the sensory and tactile defensiveness system.

I will now cover autism and moon phases.

Many people on the autism spectrum are affected by not only weather changes but our moon phases as well, especially full moons. Our moon naturally affects the earth as far as compressing the land and having a pulling effect on waves creating tides, etc. It reacts similarly to the autistic person's body by giving either the automatic joint compressions or joint pulling. For some people with autism, the joint compressions need to be lessened because the moon is already doing it for them. And, for others, their joint compressions may need to be increased because the moon is doing the opposite effect on the joints and muscles.

It actually is not too surprising, as the moon has been recently known to help regulate cycles and even help take away cancers. If it can have that effect on neurotypical people, it is sure going to affect a person with autism whose body is different all over.

22

AUTISM AND SCHEDULES, AND AUTISM AND COMMUNICATION METHODS

PECS, Word Schedules, Symbols, Dynavox boards and Sign Language are the main ways for schedules and communication usage in autism.

How to get a schedule started... (*Two ways*)

Here is one thing... take pictures with a digital camera. Then, go to your computer and type in "1" or however many inch squares you want (*check on my website, I explain different spreadsheet programs to help do that, and tell you which programs are the easiest to use*). You would then paste the digital picture in the middle of the square. Then, size them down to the size of that 1" square. You do not have to use squares if you have a general idea of how big you want them just by looking at the screen. If you do not want to do digital cameras and downloading, or you do not want to buy general pics, then just do an internet search and go to the picture link. Type in what you are looking for. In these cases, the pictures are usually small when they come up in the search engine, even sometimes exactly the right size you want. I will tell you below in steps just how much easier it is to do this if you are not experienced with a spreadsheet program on the computer. (*Using both digital camera pictures as well as search engine pictures*!)

1. As I said above, one thing to do is to go to a search engine and click on the "images" link.

2. Type in what you are looking for (*ex: Nesquik powder container*). The images will then come up and you find which one you like that best resembles it. I would like to quickly add that for later on for sizing purposes, if you can find a pic with all sides equal in length (*or close enough*), that would be better. This is so you don't have to mess around with the sizing and risk distorting the images.

3. Sometimes the image found is small enough, but sometimes too small. I recommend that you click on the link next to the 'picture icon in the search engine that says to "view picture full size."

4. Once you have the picture full size on your screen, copy it with your mouse.

5. Go to start on your computer and go to a word processing program.

6. Then, once your word processing program is open (*sometimes have the program open before you start searching for pictures*), right click with the mouse and select "Paste." Your picture will appear just how it looked when you copied it full size.

7. This is where it can get a little tricky, but I will help you through the rest. (*Remember, just take it slow and you might mess up a few times but who doesn't?*) NEXT, right click on the picture that you have pasted onto your document and select "format picture."

8. A box will open, and then you need to go to the tab, "Size."

9. You will see the words (*usually in blue but it can vary*) Size and Rotate, Scale, and finally Original Size. *I want you to focus on the height and width boxes under the words* ***Size and Rotate***. – In the Height box, you will see a full number and then an inch symbol. There will be two arrows (*both an up*

and down arrow after that). Take your computer mouse and highlight the original numbers, delete and change them to what you want them to be. Ex: 1" X 1" since most pec schedules have pecs that are that size. However, different sizes work better for different individuals. Some people will do better with the size being 1.00" X 1.00", 1.75" X 1.75", or 2.00" X 2.00" -Make note that if you choose to use a digital camera to make the pictures "personal," when you are setting up the size under Quality, see if you can't make it UGA. UGA is 640X480, comparable for email. You can just ask someone at a camera developing place (*like any photo shop*) to show you. It just makes it a little bit easier. I also want to quickly mention that when you are going to be putting in pictures from your camera, that you go to the word "Insert" at the top of your document page and scroll down to "Picture," then to "From File." Most people would think you would choose From Scanner or Camera, but that requires a very complicated setup, so choosing the other will bring you to a box. Just click on the drop down arrow and either G:/, or H:/, J:/, will be the one to choose. You will know if it is the right one because you will see your camera's name listed. ☺ I will mention later something called, "I pics," which is quite the artistry in PECS. The only difference is that the PEC size is not an even-by-even number. Change the first box of numbers under *Size and Rotate* to 1.00" and the second box to 0.75" if you get the mini I pics sheet. It is a pretty good set in that it has well-rounded general Pics to help get you started. The other set is a bit bigger, **so if you are getting confused at all by how to adjust the size, I suggest printing off one PEC you have from your ABA therapist, teacher, etc. and try to adjust the picture to match it – meaning you can put your sample PEC almost directly over your *printed* PEC with little extra space from your PEC on the sides. It is ok to have some of your PECS/Pics a little larger or smaller.** *Do not be fooled by the size of the pecs/pics on the screen, as when* *they are printed, what seems large on your computer screen is really not as large as it seems.*

10. To make it easier for cutting and knowing how much space to make for your PECS or pics, especially if they have a white border, click on the image. Then, right click and scroll to "Borders and Shading" on that list. Click on "Box" from the selection menu seen on the left and then I recommend choosing the color pink so that it is very distinct. However, any color is fine. Make sure you leave some space between your other pecs/pics you format on your document by using 4 spacebar clicks (*always save periodically*). Then wham, you have just created your visual schedule online before it is printed!

It usually does not actually take too long to put these schedules or communication systems together. Just print out the pecs that you sized down on your document (*or spreadsheet if you are familiar with those types of programs*). You can choose to put the PECS on a note-card, so they are harder and not so floppy. Then, laminate (*you can get laminate and a laminate machine for not too much at a local supply store*) or have a supply place do it when you have what you need for a little while anyway (*places don't charge too much for that at all like Staples, etc.*). You can even charge to have them attached with hook and loop fasteners for not much more in some of the same places! If your child is in school, then you can ask the school to do it since they have practically unlimited resources and knowledge in lamination and PECS for autism. Then, just get a folder, either just a single folder or a folder that has several slips. (*You also have the option to buy folder inserts for the pecs*) Just make sure you have lines of hook and loop fastener for the pictures to hook to. You take the soft sticky side, or loop side of the hook and loop fastener and put them in lines on the folder slips (*if you are buying them yourself*). Then, you take the rough side, or the hook side and attach it to the back of each of your laminated pec/word cards. (*If you are buying the page inserts, the strips might actually be hard hook and loop fasteners, so then just make the fastener on the back of your pecs the soft end, or the loop side*). Then, simply they will stick to the rows of hook fasteners. You either pull them on and off or take a dry erase marker and make a check when done with each of the pec/word cards. Mini schedules also help to eliminate more work than needed in some cases. Also, the site (*Check my website*), has tons of picture cards. They also carry the picture cards that talk and conversation strips. I know you may very well be overwhelmed right now. Just try the steps one at a time and take it in (*there is always a starting point*), as it can be very costly to do PECS/ Pics the other way through expensive programs. This way it will save you overall. Do not get frustrated, but just try it and print out the steps. Black out the steps that don't apply to you or that you don't need to lessen the steps! I do want to mention, the I pics are actually out of all the pics/PECS I've seen, the most realistic, artistic and to the point ones. (*At least for starters*) There are going to be other PECS you will need, like blood sugar meter, gf (gluten free)/cf (casein free), inhaler, nasal sprays, weighted belt, therapists' faces, etc. – which is where my instructions would come in. ☺ The site that you will need to go to in order to get these I pics is autismshop.com

From this site, you can get an actual I pics bracelet. This bracelet is wonderful for emergency personnel (*alerting*), great for reinforcing good behavior, and many other uses. For me, I have one and have found that I use it for reminding me of what I can do to eliminate anxiety. I can slip in a pic of swinging, a particular fidget, gum, weighted vest, etc. My service dog can just nudge it to remind me to look at it, but it's pretty easy when you are starting to get upset to see it and notice that it is there. It is helping tremendously!

So, are the PECS schedules and strips good even if the person is verbal?

Absolutely! I even have mini schedules all over the house, just to remember to take my medicines, eat and use the bathroom, from showering to brushing my teeth, to using my meds, flossing, etc. This is especially helpful when autistic people start to get stuck on things. I am completely lost and can be really stressed sometimes when I do not have a schedule. Before schedules, I have forgotten to eat until dinnertime. I have also forgotten to brush my teeth (*which people with autism have dental problems because the enamel is little, due to dairy sensitivities as well as deficiency in certain bodily chemicals which is another issue*) and this could actually go on well for a week if I didn't have a schedule. Schedules can help make an autistic person very happy! And, that is a good thing. I want to add that sometimes I almost wish I were still in school because the routine was so solid. (*Well, in a nice school*)

So if verbal, PECS or Words?

I have tried to switch to word schedules but they do not work for my brain for some reason very well. In addition, it can be that if it is all words, it is so much to look at and then the person gets overwhelmed. Pictures are concrete and to the point. Schedules that have pictures can work a little while later with chronological age. In addition, for me it is not just generalized PEC program pictures or it almost gets me confused in a way, so it depends on the person. Some people can just go with the generic pictures and for others it has to be "their" toothbrush, etc. As I said before, you can search for their exact toothbrush online so you do not have to take a digital photo. If you take digital photos, just follow the process I stated earlier for word and PECS schedules. It is also for me, "my" toothbrush, because since autistic people are very visual, it is reinforcing. I do have some generalized ones, but they are very different from the other PECS pictures I have seen. Therefore, it is a mixture. Also, as I mentioned before, there are PECS that you just are not going to find through several different programs, like a specific blood sugar meter or a specific therapist's face. If your child is old enough and can handle a digital camera, you can actually involve them in the process, which makes the reinforcing even greater! It can also give a sense of pride. ☺

I want to add a good advice tip for the PECS cards. At times, to help understanding that a PEC card is for something else, you can take that PEC card and use a hook and loop fastener to attach it directly onto the object it represents.

Dynavox/Dynomites are devices that have programmed responses or pictures and sometimes keyboards that use an automated voice. When pressed, they will speak for the person with autism. I am adding a video of the Dynavox in use (*It is also used in one of the videos I have under my animal assisted therapy page on my website – the Dynomite*) so be sure and check it out. It is very inspiring!

Sign Language!

A site that I have both found and used to my advantage is (*check my website under the appropriate topic*). It is a visual dictionary of sign language, ranging from child sign language (*aka baby sign language*), which is the best one to start with, all the way to the full dictionary.

Signing is good for the hands for sensory integration dysfunction (*typical in autism dx*) as well and fine motor skills. It gives great input and helps promote speech. It also encourages engagement of social cues and relationship statutes.

When I was in school and there was a sign language interpreter, I would watch them and try to block out the voices of the people such as the teacher. This is because lots of people on the autism spectrum might have slow brain processing. They may also have compromised mental vocabulary, explained so as thinking in pictures. Sign language is so very visual to the eye and the mind. I would fall way behind in auditory speech (*because of the slow brain processing*) but in the sign language, I was seeing a picture like I would see on a PICS or PECS program. I have on my site a speech program and a PECS program listed that is very helpful for comprehension. It also helps with learning to process quicker with the human body's extremely complex brain.

When you see the visual dictionary, many of the signs are self-explanatory as to why they use "that sign" for "that object." However, some of the signs leave people confused as to why that sign was chosen. Some autistic people will catch on quickly to most of them (*as typical people will*) but some will need to be shown for reinforcement as well as so they see and understand language better. So, you may take a sign and show what it means/why; like stroking a cat's whiskers and then showing the sign (*you are stroking whiskers by the nose*). Or, take a ball and put the hands around it. Let go so the sign is there to show the ball is "round." One other example to help you out is with tree or leaf. You can take a tree or leaf and spread the leaves out to match the five fingers. Let those leaves flow in the wind and then take the leaf in hand and do the sign while saying the word. This will take time but it can catch on quick, believe me! I will describe a few of the more complex signs that you may encounter and then wonder, "Why did they choose to use that sign for that word? What is the connection between the two?" I will list them below but only after explaining first what you should know as a person learning the language. The sign deaf (*see it on the visual dictionary*) was used because when deafness and the language was first started, a young boy had fallen into a fire and it scorched him from the ear area to the chin area. It caused him to lose his hearing and in happenstance, he started using signs that other children who were born in that area (*when deafness was first developing as a hurdle or rather disability, whichever you prefer*) started to learn. That is why the symbol is the letter "d" with your hand to represent "deaf," and the going from your ear to chin is to represent the beginning of the language, or rather how it came about through the scarification causing the boy's deafness.

1. Cracker: They use hitting the elbow as the sign because that area is hard. A cracker is hard and makes a noise when hit (*when you hit your elbow you will hear a noise and they summon that to crunch noise*).

2. Old or age: They use the sign to represent getting older by showing a beard is growing.

3. Also: They decided on using the sign to show that there is an object (*or person*) that joins another, so they will "ALSO… the root meaning = join together."

4. There are other signs, and sometimes you need to come up with your own stories, even if they are not the real ones as I have given you here (*the explanations behind specific signs*). Memorization is actually a game when you think about it! You remind yourself of something by connecting it with something significant.

During utilizing sign in hopes of helping conversation, I would say in a happy voice, the word you are trying to teach while SIGNING IT. Then, immediately show and give them the object you are teaching them. Some autistic kids or people need signing or symbols before the PECS start.

I am uploading a video of me using sign language with captions on autism and the benefits of sign as well, so be sure and check it out!

Also, check my site for PECS/word schedule examples I will be uploading and a video. There will be one as well of sign language and PECS being utilized in autism. Some of them will be directly from me. When teaching speech or when teaching the child to respond to speech directed at them, here is a tip:

1. Try hands on. Say his/her name and then persuade them to look at you using an autism technique they use or sounds. Then, give them a reward. Repetition is key in this process as is every other process in teaching and/or therapy when it comes to autism.

I want to add here that not being able to say something does not mean you do not want to express something meaningful to someone or the world!

23

AUTISM, SAVANTISM AND MEMORIZATION

Just to start out, this is a very fun and intriguing chapter as I was told by people previewing it. So, I hope you find it that way as well!

Memorization of complete movies and books is more of a savant characteristic. However, remembering commercials and remembering some lines of things from movies is common in autism.

For starters:

Repeating things from movies and shows is always fun for autistic people. We tend to do that because it is an "order" thing, as we know what is going to happen when and actually find it even more fun when people are amused by it. Autistic people can memorize any line for you and act it out with perfect voice imitation (*only a handful can do perfect voice imitation*), but when it comes to tying your shoes or remembering to take a pill, it seems like it does not make sense why then we cannot remember those things.

Savantism and my experiences with it are given below (*the psychologists and professors stare at me in awe with both utter amazement and looking so very confused, LOL*):

I, myself, have always been capable of remembering things that have amazed the doctors and psychologists. Now that I know that, it is fun for me to amaze the doctors and psychologists! However, it really does drive me crazy sometimes. This is because I can't stop remembering things, even if I desperately want to. The psychologists asked me tons and tons of letters and numbers and then I had to turn around and repeat them in numerical and chronological order. I got it completely correct according to their paper. The psychologist after that said he had to stop and go onto another test. This test was common to be passed by a younger child, but that one I could not understand. It seems confusing that I can watch a movie or read a book and remember the lines (*I had watched Pirates of the Caribbean and repeated the whole movie for my special education teacher over the course of three lunch days*), remember number and letters, but I can't remember to take my pills. Or, I forget to call someone back in a minute. I want to add here about when I repeated a whole movie for my special education teacher over three lunch days. He had said in a joking way, "Ok, now rewind or fast forward ___ of time." Believable or not, some autistic individuals watching and listening to the movie will use their peripheral vision to then look at the timer on the recorder and remember it. It is an unconscious but happens. I was able to do a few, as I did not look at it all the time. However, I sometimes do it more now. Nevertheless, I think that is also because I get stuck on things, especially because I become stuck on numbers the most. Sometimes it makes me saddened because I cannot stop having all these thoughts going through my head. So, it is actually overwhelming to say the least sometimes. The psychologists (*testing me*) had showed me cards (*they had symbols on them and very specific rows, etc.*) and would show them to me for a few seconds. There were thirty of the very specific patterned cards. Then, they put them back in their case. They said, "In a couple hours we'll see if you can try and remember even one of the cards. Don't worry if you can't, as there are only a handful of people that can do it, so it is not a problem." When they got to that point, they said I could try and draw them and in the order I saw them. They wanted to see if I could draw an exact replica of the symbolized picture from each of the cards shown earlier. For my brain, and many autistic people's brains, the file comes up slower. It was taking me about thirty seconds and they smiled and said, "Never mind, (*they laughed too*) most people can't; only a few can and you don't have to do that!" (*These were different testers than the testers who I did all the numbers in chronological and numerical order, so they weren't aware I had this kind of ability*) I said, "No, I want to..." Therefore, the tester went to get another psychologist for the next part of their testing. I had drawn about 14 of the total 30 cards specifically after the visual file finally came up for me. I finished them with absolute accuracy and after they pulled out the cards, they stared at me and then at each other. They then asked me how I did it. I said, "I don't know." Therefore, that was pretty funny, but that is how my brain works. It is as if I CAN'T forget them. When I was younger, I was almost always sitting at the microwave and counting, figuring out how the time left or the numbers would correlate to each other. I'd try and figure out how many times the numbers 1-2-3 would come out even if I kept doing sequences of 3. I was only five then and keep doing that stuff. It was really what I did for fun along with taking my toys and "fixing" them. I would also look at the time on a clock and figure out how the multiplication could add up to equal the full amount of numbers shown, as well as the entire equation of it. I know that probably seems rather tiring and not fun for you guys, but for me it is exciting! Other people in my support group are amazed, and the psychologist said that what goes on with me is a savant thing. He also said that every autistic person has something they can do well at and surpass others with it remarkably!

When I was going to school, I was at a math-tutoring place where very well-known professors of their field such as calculus and math related to space and physics/CSI expertise were playing a game. They were trying to figure out the pattern and placing of numbers in a very large box of numbers. They had been trying for literally two and a half hours. I came in and then after a couple minutes I went up to the board and started plugging in the numbers in a different color so you could see the plugged in ones. They checked with the person who found this in a book and it was correct. For me, I saw it as a picture or design, as it was like reading English (*although I do have bad comprehension so that is not the best analogy, but it was similar in that way*). A good representation of this is in the movie, Mercury Rising. I have this example on my movie trailers page on my site. It is somewhat similar in nature to the one that I showed the tutors how to read because of the numbers. However, in the motion picture, there aren't any missing numbers because it is an actual code.

ALSO, my abilities in more numbers is coming soon, as well as other personal and extremely detailed memories of things that I told my parents who then talked to other family members and found out I was completely right! However, I will only mention a couple occasions in this chapter just a little later on.

My parents really like watching the show NUMB3RS and they laugh at all the stuff saying things to each other like, "Wow, could you even imagine understanding even a second of what they're saying?" In addition, I joke back, but the truth is I actually understand lots of what they are saying on the show and smile inside. When I was in college for a bit, I would go to the math-tutoring center and help out the tutors. They would help me find the words or actions to get out what I knew inside about these science and mathematically related concepts. It was more than difficult because of the autism trying to block it all from being able to be explained. So many times in autism, it is not uncommon for an actual math disability to be put down in an IEP. It is a lot of times because of this block but also because some of the actual formulas for some reason cannot be remembered by the autistic brain (*strangely enough other things can that are much more complex but that again is a part of the mystery of it all*). Also, the understanding of these concepts can be bottled up in the brain for many years and all of a sudden start to be understood through unique methods of teaching. I explain these methods in the chapter on teaching.

Here is an example (*much easier and quicker than the others for me are*) of a math problem that I like solving. I have broken down this one to make it a lot simpler and I actually make them, which is even more fun!

13251809_132

35581111222

_32518_1_2

_3121_ _1328922222

_1_ 312112_1113_ _1_ _ _ _2

Because of my memory, my friends hate playing games that require memorizing the location of matching images or similar type games. They usually always lose and have about two times to turn over the cards. I have actually resorted to buying several decks of cards and making it so that I have to pick up 10 of the same cards that match before I can call it a "pair." It is so much fun and does take a little while but actually not too long, as my memory is very photographic.

I actually make up Sudoku puzzles as well. I have a handheld game of it and a book. I am doing hundred column games. It is really quite interesting and I feel helps my brain keep thinking. One of the best games ever!

I would also like to add that I have a love for deciphering vocabulary and language derivatives/roots. I would be in school and be the last one standing for "spelling bee" things. It is very important to me, and believe it or not, I love reading the dictionary. Similar to this and sort of similar to word scrambles and other word games, I will take a list of names and figure out a way to connect them all together by the first letter of each name. Also, I make it without any of the names running into each other. LOL, I used to do it sometimes on the old TV when the dust collected on it so you could write on there! Again, it is just something that I find fun for some reason and I guess it makes me feel happy because I feel unique. I can feel I guess a little proud about something, because although I may not be able to do things that others can that should seem easy, I can do these other things to make up for it. (*I would like to bring up here that many people with autism really do not like hearing compliments and that is still an autism mystery*). In addition, what I have found is that you can take a foreign language that has letters and look at the different ways they are put together, how they sound within harmony and then you can sometimes decipher what a particular or certain sentence indicates. For languages such as Japanese, Chinese, and Korean it is more complicated, but that makes it even more fun. It is like a puzzle, because you can find symbols that are constantly used in certain parts or at ends of sentences. You can then try and figure out what a word might be to make sense for a sentence with a specific number, word or symbol. That is the other way of decoding, which takes obviously much longer but I think is a blast. I just love contours/elements!

Savantism can bring as well as all the autism spectrum disorders, many things like talents but also indecision. So, it is like a gift and a bother too at times.

Before going onto the next chapter, I am going to tell you that my memory is very precise and I have been able to tell my parents and relatives entire talks that were talked back when I was just even four and five years old. Aside from that, I have described how places have looked, how many stairs someone had at these ages, where everyone sat in restaurants and events, what they ordered, etc. Of course, I will not go into the details of all of that, as that is just family stuff and would not be very interesting to you. However, I'll give you a few small things to show you in words a small bit of what I can remember including now.

I remember for instance, when I was four years old in preschool, that we used to go into the gym that was a gray floor with squares. To get into the gym there were 10 brown plastic feet mats leading into there from the room that was connected. I also remember that after school, we would go down a hallway to the left and then take another left until we got to the doors (*like in a mall sort of*). There were three candy machines. We would stop to put in the quarters and I would sort the banana and cherry candies.

Also, I remember a while before moving, we went to dinner when I was about five years old to a neighbor's house for dinner. I remember his table seemed like oak with unique chairs. We had hamburgers with ketchup and pickles. I could go on but then I also remembered that there were three wood steps going down straight across from the table coming into the front door. There was a deer head on the wall and old-fashioned television below. It was like a hunting type of room. My parents were amazed, as years later we went back to visit. We talked about our last dinner there before the move. I was only five during this event. I verbalized this maybe around 13 years old. This is around the same time I talked about the other things I talked about earlier from preschool years. Sure enough, it was all exact. My aunt was laughing over all of this as well as my mother and myself, my cousins, etc. As part of it, we enjoy sorting those banana and cherry candies even more now!

The things I remember are so specific most of the time and at young ages that my parents are always shocked and feel the need to share it with the family that was involved. I'm glad I can remember these things because I would not want to forget them – memories!

It has been reviewed that some people who have these abilities can start to lose them when they go through therapies and start focusing more on the world around them, but it depends on the change in the brain from therapies and the level.

At this point, last but not least, my friend and I a few years back had went to a fast food restaurant that had little jukeboxes at the tables. We had glanced through them real fast and were eating. A song came on, I said J3, etc., and so then my friend said, "Yeah right!" Then, when she saw I was right, the couple people who were taking a break (*who were working there*) thought it was cool. My friend and they were laughing and quizzing me. I will NEVER forget it! Did you get the NEVER part as part of this chapter? Thought it went along pretty funny! ☺

I hope this chapter was interesting to you as well as wonderful, because you can see the interesting things of autism.

24

What it feels like to be autistic

If you go to my website, I have created a slide with specific colors and visuals to help you understand each thing I mean here. However, I am going to give you the text of my statements that I included in the slide. I did this project for a psychology class. I want you to know something in advance. Sometimes and many times, when you are autistic, you also have a ton of peace and other times can feel almost like a typical person. (*I think*) However, to help explain how our body responds to vast amounts of input, this is how I describe it:

1. Feels like 20 cologne smells and your head is likely to explode
2. 50 children running around you
3. Nails going down a blackboard
4. Your phobia surrounding you
5. Your brain and sensory feels like it is not comprehensible and cannot be decoded, similar to the squiggly lines on cans or packages that are scanned in stores
6. Seeing all detail of everything, even if you don't want to
7. Lights shining brightly everywhere
8. Just plain confused and feel exhausted from trying to be like others

Sometimes when I am overstimulated, it is like the commercials they make for ADHD (*where there are several things going through the mind at once*). It really is like that in autism when overwhelmed immensely. Sometimes this creates the head to shake back and forth, crying as well as hands being put over the ears. It can be like your phobia in 3-D coming at you and you can't shut your eyes.

It is at this point that I am going to bring up a couple stories that I said I would in a previous chapter. I was about 7 years old and me and my family, along with my grandparents, were at Universal. We were watching a show with pirates in water with fire. There were children about three and four, even younger next to me, and I kept closing my eyes and ears. I was crying and really getting overstimulated. My mom had to take me away and out of the bleachers because of the autism. Then, I sat down where there weren't a lot of people around. A couple members of my family stayed and finished watching. One other person left to come and sit with my mom while I sat there recouping from everything. (*They didn't know I was autistic yet.*) At this point, I would like to bring up that recouping can lead teachers to believe that their autistic students are using tactile stimulation objects like shoestrings or hook and loop fasteners, etc. to get stimulation, but really what it really could be called more accurately would be a tactile stimulation desensitizer. I am saying this because we, (*the autistic people*) in these specific situations (*situations that have a lot of stimuli*) are using these tools or objects not exactly to get more stimuli but to get a negative or opposite reaction from that object. Therefore, it can push away all the other stimulus that are bodies, brains and systems are getting. This one teacher when I told her about these objects I always carried around, connected it to why her student was clutching onto them during times like P.E. I just had to add something here that she emailed back to me that made me laugh. She said to me, "I just grew some more dendrites!" Therefore, in some situations, these objects are to help us feel our bodies. But, many times in overstimulation cases, they are actually used for the reverse effect. So, you should be careful when thinking of taking away a stimulation object because of worrying that the situation is already stimulating, thinking that the object then is only going to add more. It is doing the opposite reaction, and a good reaction. It is like a negative magnetic force that produces only good and in the result produces positive and balanced energy (*or stimulation to the senses*).

There are times in autism where there is such peace. I rather explained that somewhat in the chapter about autism and love, but want to reiterate that there are times where it is so beautiful as well aside from all these struggles. Also, a great thing about autism is honesty and animal connections, etc. In addition, having the savant abilities can be fun. The other thing I would like to mention is that people with autism can many times be easily pleased and not so highly disappointed. How many people who are neurotypical that you know would be highly entertained and joyous by being given a shoelace or a plastic spoon, a blade of grass, bark off a tree or a wrapper? It is good that a person with autism can be pacified and completely joyous without having to have really big things. Having the ability to be as joyous over the smallest things as much as other things are I think wonderful both for the person with autism and of course the parents. (*They don't have to do much to please!* ☺) There really are great things about autism and the love that goes along with it. Be calm and relieved to know that aside from these struggles that come along with autism, they can not only be sometimes overcome or helped, but there are so many other sides that I wouldn't trade for anything.

In an upcoming chapter about autism and quiet thinking, I am going to bring light to a certain subject. I feel it also well to bring a taste of it up in this chapter as well, as it would tie in with this chapter.

Many parents will say, "I wonder what my autistic child is seeing/ thinking." "I know that they say it runs tons of times faster than the typical person's brain." My response to that in part is that the autistic person is thinking a lot in different terms. Sometimes they are not exactly thinking verbal thoughts but just seeing scenarios, seeing details of things, and they are simply looking and visualizing.

I want to bring up here that there has been work on a machine to help tell what children or adults who cannot talk or move are thinking about (*the machine would show a strip of designs or pictures that match what the brain is visualizing but cannot say*). This is an amazing work and could help explain, help verbalize, or show what specific individuals are feeling and visualizing. Some of these individuals may be brain injury victims. However, of course, these people may not always want that because they want freedom to think for just themselves at times and of course keep their thoughts secret.

Sometimes I cannot really explain all of what I am trying to say, and cannot even get it out through writing, so it is very frustrating. On my website, I have two videos that really help me to show you what it feels like, since the videos speak for me when I cannot get the words to come. Thank God for those who can help get the extra things out that I cannot. Without my learning and therapies, I would not have been able to write this book and even explain half of what I have in here. And believe me, there is so much more I wish I could explain, but I just don't know how to get it out. **Alternatively, it simply is stuck in my brain.**

25

Autism - Why some want a cure and others are against it

There are so many parents who desperately want a cure, some people who are pretty high functioning and are advocating that autism is not a disease but a way of life, some autistic people who vouch for a cure partially, and then there are some who really just don't know and don't want to even think about it.

For me, I am the person who vouches for a cure partially. I will explain that in just a moment.

Yes, every person with an autism spectrum disorder is special and unique. There are hardships and struggles, and then there are strengths and desirable traits.

I think the most people that are concerned to find a cure and who advocate for one are those who have to deal with the severe sensory issues that do cause pain, vestibular and balance issues, speech difficulties (*not only apraxia but also literal issues, etc.*) and swallowing and gagging issues associated with sleep problems. There are also those who are higher functioning but do have difficulties socially. Many wish they didn't have those problems because they feel they don't really belong to a group and feel like they want to be able to have some of the things that neurotypicals have. There are those who are high functioning and do not have to deal with as many of those problems, or not as much. They have great talents that they use proudly (*and that's good*) and want to be looked at as unique only and not as disabled.

I am partially for a cure. I myself want a cure, but then do not want a cure. One reason is because I am afraid of change (*part of the struggles with autism*), I wouldn't want to give up those savant abilities and I would not want to change the beauties that I explain in this book (*about autism including for example being able to connect with animals in a very special way*). I also would be a different person. ☹

Yet, I would also want a cure because of all those things that I mentioned before that I have to deal with. I believe autism is just one of those things: difficult with beauties. I would take a cure if it would take away the struggles and leave the beauties, not to mention sometimes parts of pain resistance (*where you are numb to pain sometimes physically – When I had braces, when others would be in pain from the tightening of the ties, I asked if they could tighten mine more because I couldn't feel much, but they said they couldn't because it could break my teeth*). The underside to this is the risk of the person not knowing if they are hurt, so they have to be checked periodically and have devices to help for temperature regulation. I had one night been coming down the stairs with my cat. I tripped and started to fall. I didn't want her to get hurt, so I fell with my arms holding her and then, releasing her just in time for my elbows to bang into the hard floor. I got up afterwards like nothing had happened to me and my mom stared at me as if she was in dis-belief. It is like it would not occur to me at times when I could be hurt. That is the underside to partial good part of having autism – the being numb to pain sometimes physically.

I guess you just have to monitor!

26

AUTISM AND THE VISUAL FIELD

Eye contact issues can be really difficult, especially if you are higher functioning and someone might not know you are autistic. They become quickly offended that you are not looking at them. I had a tutor once get very offended I was not looking at her. She got very angry and demanded that I was not respecting her while she was showing me something. I went right to a psych. teacher I knew. He said to me, in a joking but serious way, "You could have said that you are part of a culture that actually sees direct eye contact as being disrespectful. There are many cultures and part of belief systems that think eye contact at certain times is not warranted or not wanted." This is very, very true. I decided I could use this as a tool if a person becomes that way again. There are also groups that can teach an autistic person ways to get around these situations by having them look at a person's hair or something near the person. This is so that you can sort of be making contact with the person and the person may even think you are looking at their eyes, but you are really just compensating!

Why is eye contact so hard for autistic people?

There are usually two reasons. I will list below the reasons.

1. If you are staring at the eyes, you can become lost in the conversation or completely forget what you are saying because the brain cannot handle the two things together.

2. It can be scary to the person and the eyes can become so pertinent as they are bright. In war, they are referred to as the body's very strong part of the face as they become threatening.

I want you who are reading this to think over what I am going to say next.

A person with an autism disability can have an extraordinarily hard time looking a person in the eyes (*that would include me*), but can stare into an animal's eyes for a long time and seem to communicate with them. (*That would also include me*) Is that not interesting? It just goes to show the nature of the autistic person and the similar nature to an animal. That is my positive observation on my thinking that Autistic people are calmer and more like an animal in lots of ways.

The other part of eye contact I am going to bring up has to do with autism and staring. This does not necessarily have to do with the autistic person staring at another person's eyes, but rather at the other parts of their body. This can be very common and a red flag for as well. My brothers would put up cereal boxes or objects in front of them because I would literally be staring at them for long amounts of time. I would like to re-mention that I mentioned in the chapter about dealing with the staring. Again, if you yourself have an autistic child and you want to yell at that person who is staring at your child, they might actually be autistic. The reason for this is that the autistic person becomes fixated on a movement that the person is doing, they think something is unique about them, etc. and become intrigued. We are not trying to do this to be rude or make someone uncomfortable. It just happens and some of the time it's almost like a temporal seizure because the brain just stops thinking about what it's doing connected through the visual field and just hangs there. That is why I especially wanted to bring this part up because for autistic people that is a red flag tendency and is a part of the spectrum. Therefore, I thought it was imperative to bring up in this chapter about contact with the eyes and autism.

27

AUTISM AND TRENDS

I am going to cover three areas in this chapter. The first area I am going to cover is basically about autistic people and their role/ understanding of trends, the second referring to autistic people befriending several people despite their age, and the third category that talks about the autistic person's age and where they feel they really belong in that status.

Many people with autism do not look at age and think of it in terms of sociality. Similarly, many autistic people do not look at fads as well as "age standards" that the larger group of neurotypical people would judge. For instance, many people on the autism spectrum will wear clothes that society would look at and say, "A much younger child should be wearing that and she should be out growing that type of clothes and hair accessories long by now." It can cause problems such as others making fun. However, well since I am autistic maybe it's just my thought on it, but I think about the fact that certain color clothing or designs are thought of as outdated, inappropriate, or need to be changed. I wonder who decided they became coded colors or coded patterns. Sometimes things in our society (*well, most things*) change and that's just how they happen. How neurotypical people view change is influenced greatly by publicity, whereas the individual with autism many times does not see this and just wants to feel happy. The autistic person's view many times is, "People say to do things that make you happy, so if this is innocent and makes me happy, then I don't understand why you would change your mind now." That is obviously, as you can tell, part our literal and simplistic thinking associated with autism. Also, the change in the world happens through formation of character. For example, some things would be fine and then were worn or were present at the time of a non-innocent affair that took place. Then, impulsively, they are viewed as inappropriate unfortunately. In addition, as you can see, the autistic brain does not view things that way and just views them as they are at that specific time; like Mother Nature. Our world is viewed (*hence the title of my book*) as a prism and viewed as slices of vision. Whatever slice is next is what is seen, so we do not necessarily see how the neurotypical world sees trends. I hope this is making some sense, as it is coming from me and I am autistic. I am trying hard to explain it as well as I can for those of you who are not.

I am now going to talk just briefly as an add on about autistic people befriending several people despite their age. Many autistic people also do not view the figures of a person, but rather the interest of the person. They may be older but enjoy spending quality time with a much younger child because the younger child has traits or loves to do "younger trend" things. So, the autistic person doesn't mind at all the number associated with the person as it doesn't mean much. Comparable, an autistic person may thoroughly enjoy the company of an older person whose calculated thoughts and wisdom excites the autistic person's part of their complex mind. For many younger autistic students, you may many and most of the times find that he/she is eating and spending all their recesses and lunches with their teachers. They may enjoy helping in the secretary's offices too. In turn, most the time, it is great because the adults enjoy the autistic person's company and help as well. In college, I was even then eating with the instructors and helping out by stamping things in offices or sorting music. Part three is below and similar to this last part because it deals with age. However, it is deepened because it deals with personal age.

Many times a person with autism when they are younger, feels like they are a young adult stuck in a child's body. Then, when they get to that young adult age, they feel as if they are stuck in that body when they are meant to be in a younger child's body but also in an older being's body. As time goes on, and as they get older, they can feel relief that they are getting into the older stage but still feel stuck in the wrong body. It is a hard feeling and one that is I think one of the hardest mental things to deal with emotionally; feeling like you are stuck in another body you are not supposed to be in. It is like that throughout life every day for many people with autism who have that intellectual thought process (*many do*) and even for those who have underlying disabilities that restrict that area of the brain. This is because the emotions take over the role-play of intellectualness.

I do hope this chapter has enlightened you as you have read it and I really hope it gives you a better understanding of our view of things. In return, I hope it gives you a better understanding of my partial reasoning of the title I no doubt chose for this book!

28

AUTISM AND SLEEP

Many parents of children with autism can relate to this chapter whole-heartedly and who could go on with stories relating to their children and sleep issues. (*As well as other chapters in this book*)

Many people with autism have trouble falling asleep. Therefore, there are lots of methods being used such as medicines like Melatonin, etc.

However, medicine is not always the answer, and you can use natural resources instead of resorting quickly to chemical ones.

I have had trouble falling asleep as well as waking up during the nighttime. I will go into more detail on this as well as night terrors/ day terrors (*believe it or not, there are day terrors*) after I give you some tips on what has helped me and can hopefully help you.

1. Listening to soft music (*autism-calming CDs are extra nice*) like nature sounds or water is really good. There are specific CDs that can be made (*like the auditory process CDs and listening therapy*) that are for your child's specific brain wave patterns to help correct them by way of computerized technology. You would go to a neurologist for this but also want to consult a person who is an autism expert. They can lead you in the right direction for this as it is more complex. They would test your

child's brain waves while at night or in the dark and save the brain waves on file. They would then come up with a correct form brain wave that would either merge or counteract his/her original reading to help normalize the waves for sleeping or other times. This would depend on what showed up in their test results. It can cost a bit, but is well worth it. You can find some similar CDs but they of course are not direct brain waves to your child's brain. You can find these ones at places like Barnes and Noble.

2. Swinging in my net swing with the light low for a little while right before getting into bed really helps.
3. The weighted blanket across lower end of legs, like ankles area can really help. (*Now it's my service dog!*) I will wake up but her weight on me just puts me right back to sleep.
4. Believe it or not, sometimes having your child's favorite episode of something on "really soft" on sleep mode can help them fall asleep quickly. Some autistic people have problems if they don't have some type of stimulus when going to bed because their mind and body is too relaxed. It's a hard concept, but true.

I now want to go into the subject of autism and sleeping little, but seeming to have energy as if there was adequate sleep.

When I was younger, up until about tenth grade, I would fall asleep for about 1 ½ to 2 hours and my system would automatically wake up (*another relation to many animals; the automatic alarm system*). I would go sit and lean on the wall next to the railing for a few hours, moderately awake and sometimes stimming, until my dad's alarm would go off for him to go to work. (*Which was pretty early in the morning*) I would then run back into my bedroom and sleep for a couple to few more hours until it was time for me to get up and get ready for school. I did not have that much sleep and yet I had energy to get through every day as if I had adequate sleep.

I used to have night terrors and even like a day terror sometimes. Terrors are common in autism. The brain is fired a thrill reaction from the overarousal of the Central Nervous System t which causes the night terror. It is the chemical imbalances and is agitated by the stimulation and stress running in the brain all day, which also causes in autism more fatigue with sleep issues and behavior. Then, it runs quickly through at night while the muscles in the body relax. And as a result, there are electromagnetic stimulatory particles (*red e-particles*) and then "bang" you get the night terrors. During the terrors, it will seem like your child doesn't know who you are or anything for that matter, seeming to be making no sense. The eyes may be open but the brain is unable to register it because of the disturbance so he/she is not able to tell who you are or what environment they are in. The scenarios like above can likewise begin to happen to the child during the day in disabilities such as ASD from too much stimulation or change as well. You don't know what is going on with your child with autism because they are acting as if they are in a nightmare. It is because the brain is having what is depicted as a day terror.

Sometimes, as the person with autism gets older, that scenario of little sleep but energy can turn into "Can't wake up!"

Sometimes, with good schedule practicing, the individual with an autism disorder can have an automatic alarm clock so they will wake up at the same time (*even literally to the minutes and even seconds*) similar to the ability animals have – another connection! (*This also goes for bowel movements and urination, as the person with autism can develop a pattern at which time they "HAVE TO" go use the restroom at a certain time every day. This can cause confusion for teachers who don't understand that concept of autism*)

I hope that these tips can be helpful to you and your child with autism.

I also sincerely hope that the issues I brought to light after the tips have helped you to know that these things you may be going through are not unusual for autism.

29

AUTISM AND FINANCIAL SUPPORT

Since autism causes such vast amounts of issues that require therapy to help, the money in every autistic family's banks seem to disappear a whole lot quicker. Some families have even been told that their house is going to be repossessed because many of the therapies are not covered by insurance. Then there are the medications and the supplements that are used to help with autism and the immune system, dental hygiene, and lack of calcium. All of these things seem to take over the money for food, bills, the car payments, etc. Lots of parents are not aware of at least some of the financial support they can get from the government. Social security income and social security income disability is a growing hope for those few families that know about it. (*Also insurance*) I want every family to know about it, so that is why I am covering it in this book. My parents had been talking with my good psychologist about how autism therapies take all the money and how difficult it is to live on an autism affected budget. The psychologist brought up SSI. My parents had never thought of that, as many people have viewed it for those who are physically disabled or families of Veterans. I had been only diagnosed with Asperger's when we were granted SSI. I had not even been diagnosed with the label of autism yet and was still approved to obtain SSI for my disability. It helps my parents with the therapies, medications and the other normal family expenses. Sometimes it is not much considering the costs of everything, but it definitely helps.

A brother of an autistic boy has written a letter to some place up high in the rank about insurance covering the expenses of autism and related disorders towards the new presidency. So, hopefully it will be taken into consideration soon and acted upon by the hierarchy as it should be.

I also wanted to go ahead and mention here the importance of insurance for these therapies, as they are desperately needed because autism is a bunch of disorders combined together, hence the word Spectrum. The syndrome autism includes disorders such as ADD/ADHD, some Schizotypal personality disorders, Tic disorders, OCD, SID, and more.

I would also like to add that SSI should be warranted as well, because people who are typical are usually out starting a job, whereas the autism can hold back years the person with autism getting a job. This shows that disability income and insurance is needed there too. This area has to be looked at as well as overall income.

Also, you as the parents or caregivers have been paying this department out of your pay checks, so when you need what you have been paying for, you deserve it.

There are also other services available to your families aside from SSI and SSI disability. A couple of these services include DDD, or the Department of Developmental Disabilities as well as Rolling Access Funds. These two resources I just mentioned that you have to qualify for (*and can almost automatically qualify for through the referral of a clinic or your psych who did the dx*) can help with respite care, therapy services in the home and outside, camps, technology, housing related fees, and the list goes on. In some states and for many places, in order to qualify for services, your child must have been dx before the age of 18 and must have the word "autism" in the dx; not just PDD, PDD-NOS, or Asperger's. They equally deserve those services, as at times, they can be almost the same and all require intervention. There are ways to work around that qualification rule (*by your psych referring you there and writing a note as to why you should receive the services, videos or other sources such as school behavior and adaptations made during it, etc.*) because many people with an ASD go undiagnosed too long or they have a label that doesn't have the word "autism." However, it is included under the umbrella. When you read in my chapter about the DSM-IV-TR boards litigation decision, hopefully that will help as well to help these important services become available to you when you really need them but there are things stopping you from getting them. Do not just let things be as they are told to you if it does not seem right, because you can use what I said before to help you get around things stopping you from getting what you deserve. They are there to help the person with autism and their entire family as well. The rolling funds can be acquired before DDD if needed and you just need to call your local chapter of autism services for help to get there! They can help you when you feel like there is not a lot of hope or money and time is limited during times that the autistic person really needs those crucial services and amounts of time (*as well as the family so they don't have to quit their jobs completely or in cases work more than one job, etc.*) I hope that has helped you and you can look up online these services, and as I said before, talk to your psych who got you/your child diagnosis.

There is support; you just have to know where to find it and how to get it!

30

AUTISM AND THE WORKPLACE

In this chapter, there are recommended jobs that can be great for people with autism. I also have set foundations and tips for both the person with autism and the person/s they are working for or with.

Having autism can lead down a path that lots of neurotypicals would find really hard to work in. For example, many people with autism like repetition and therefore could be very successful and even really enjoy assembly line type work. Most neurotypicals would end up finding themselves constantly looking at their watches and soon after probably turning in their notices. For a person with autism, though, (*and I have talked with parents whose children with autism do this type of work*) many of us would be excited and like the repetition, especially if it is something mechanical or has a texture or smell that would appeal to the senses. Some autistic people have thought down upon some of those jobs that they are really great at because they start wishing they could do other jobs like being a pilot or astronaut, etc. However, the autistic person's job is very important. I always say, "How would these people ever get to space if they didn't have people working in a factory or assembly line preparing and/or packaging up special food, the different parts for their space compatible spaceship, etc.? And, it is really a great service because you may even be packaging up food for people who need it, getting supplies together for doctors, etc. You are basically helping these jobs that you say you wish you could do.

You are a part of this job, just on another end of it. You are doing something important that most people wouldn't do, and because of it, you are a great help and very important."

Then there are people on the autism spectrum that have a talent selling beautiful works of art or music that help others feel happiness too. And there are those people who help as puppy trainers or who help work in an office at places like that.

Aside from that, there are people with autism who can crack codes that are very highly important for the country. There are also people with autism who have invented things that are so great today and those who are helping along really complicated things in our ongoing history.

There are people who fix instruments and clean them. I personally know a few people with autism who do this for a job and get paid well. They also really enjoy doing their job for the directors, their instrumentalists, etc.

There are many reporters for writing in demand for important articles to inform the public and many people with autism communicate very expressively through their typing.

As well as that, there are plumbing, air conditioning, farming, and many other tedious but rewarding jobs that the autistic person could relish at because of their unique ability to persevere! They become extremely talented in their field.

The possibilities are so great and you should see the talent and unique passions the individual with autism has and do encourage it along.

Having people like your Vocational Rehabilitation place on your side may be very beneficial as well, but if the person with an autism spectrum disorder thinks they want to try a small community college, that shouldn't be pushed aside in cases. There are quite a few people with autism disorders that have even gone on to universities. There are now college and university type schools that are just for people with autism and/or similar syndromes. They are geared toward the specific autistic learning needs while incorporating other skills as well. The people there know how to teach at the right pace, what tools to use to help, and in my opinion, a graduate certificate from a school like that should be taken in great pride because it is a school that helps people achieve great things in life in a special way!

For a workplace, the person with autism should have a schedule and be offered assistance like job coaches if they need them. They should also be told several times beforehand if the person they are working for or with knows that there will be a change. Help that along as well.

I was told that if I wanted to, this guy who is the head trainer at a Petsmart, said that he understands autism really well and sees my connection with the animals is amazing and would be more than willing to hire me and help me lead training classes for a job. Sometimes the person with autism does not need much training because the passion they have for these jobs or the type of job they would be working in automatically reflects back their ability to be able to do the job. And, in my case, I would be receiving help on the job while being able to earn money myself. In return, I would also be doing a service for others alongside my service dog.

Another great thing to take note is that a recent study was done that says most neurotypicals will miss work approximately 14 days out of their year in the job, if not more. And, most of the autistic people that have been hired only miss a couple times to no times at all. Another plus is that many autistic people will stay with the same company they are working for, while a person who is neurotypical is consistent with changing companies they work for or locations every few years. That of course is not everyone, but it does show that the person with autism with/out help can be very prompt and a very worthy employer. It also goes to show that many autistic people have it in their nature to work and to please!

31

EVIDENCE NOT MANY PEOPLE EVEN KNOW WHAT AUTISM IS, EVEN WHEN IT IS IN THE NEWS

This chapter is going to be a very short one, but I had to add it to this book. This needs to be brought to the attention of others, so that the police and medical teams are aware of what autism is. It seems as if recent news on the TV should be having the effect it is trying to give: awareness. However, there are truly still many people, especially those trained to protect, which of who are unaware.

Here is my unfortunate example of what happened to me one time that completely shows this chapter's evidence.

One morning, I had gotten a threatening instant message on the computer that had also been similar to those being sent to scare people in email boxes. It ended up being a hoax, but since I am so literal, I took it seriously and got a little nervous. I called the police and they came over to see the messages on the screen. Because of my autism, I was acting different and stimming a little bit. The police officers asked me if I was just nervous. I told them, "No. I have autism." They said to me, "Oh. What do you take that for? Anxiety?" I tried to explain that it is like the movie, Rain Man sort of. The officers thought autism was a medication. Is that not just sad? That could be dangerous.

Although I can now laugh about it, it is a serious thing. That is why I have sent my site to local police departments and medical divisions. I hope that they will look and start giving required seminars to these officers and others in similar fields. I hope by writing this book, these people will use my guidelines. They should utilize these guidelines instead of just knowing what autism is and that is as far as they get.

32

AUTISM AND THANKFULNESS

I thought I'd share this with everyone in order to show you that an autistic person can be so thankful inside, although you might not see it since it is not expressed in the regular fashion. I know it seems strange in the beginning but you will understand (I hope) towards the end of this little thing I wrote. It will say in bold, "Ok readers this is the message." ☺ You cannot miss it!

To my cousin:

You have lit up my world.

When you weren't looking, I whispered, "Will you be my brother too?"

And when you were, I always thought of you as another big brother.

You were such a helpful person: When I was afraid of the butterflies on the hike we took as two families, you said, "It's ok." And you pulled me next to you and made them go away so the world wouldn't stare.

And even then… No one knew why I would scream or why I was different, but that didn't matter to you.

And when I saw you before you left for band and we were going to leave for home, I whispered in your ear, "You are the best brother ever." It was the first time I really ever said anything that I wanted to say from my heart and it's because you inspired me to be more social, even before I got into the social skills classes.

The only thing, sometimes you can't get me to shut up now! LOL. After all, the name Kriscreama didn't come from nowhere!

And you've helped others too.

And because you cared...

You showed others that it was ok I was different.

I would describe you as:

A helpful brother, a family guy, caring, husband and father figure, non-judging, joyous, helpful, thankful for all the things not materialized such as companionship family, an accepting brother, a God-sent son, and gentle.

What more could a person ask for?

The joy you have brung to everyone you know for those who showed it, and for those like me who didn't always know how to express it.

Above all, what I am trying to say is this:

Thank You

For

Just

Being You

I am honored to call you my cousin, and to me, a brother too!

Cuz Kris

First off, I hope the person who I am talking about, knows I am talking about them!

Ok readers this is the message:

Secondly, I hope this shows all the readers out there that although an autistic person may not necessarily show how grateful they are for something, they really are. They just might not know how to show it. I did this by PowerPoint to get my message out.

33

AUTISM AND REMORSE

There has been confusion when it comes to autism and why an autistic person may start laughing when something is happening that should be remorseful and saddening. The misconception is that they do not care, which is not true.

Have you ever seen a person crying and laughing at the same time? It is similar with an autistic person. When an autistic person is laughing, there are usually two reasons for this. The first would be that they do not understand that what is going on is upsetting or that someone is hurt. They may be thinking that the person is just acting funny on purpose. The second is much more complex. An autistic person's sensory system is already heightened as it is. You will see in certain religions that the people praising are laughing. In autism, when the autistic person knows that the thing they are encountering or watching is sad, their system does not want to respond in the normal way. The system is in such sadness, that it starts to cause laughing. Laughing during these times is a response that is far more remorseful. Their laughing should be looked at as one step above crying... such as great remorse. The system is so caring that it reacts very strong, just in a different way that to the neurotypical brain is brought on only by something that is hilarious and good. I hope that those who read this book will also help end the terrible rumor that an autistic person, when laughing at something that is not supposed to be laughed at, is not feeling remorse at all and does not care. It is actually the extreme opposite or just plain misunderstanding.

34

AUTISM, ITS PUZZLE SYMBOL AND WHAT OTHER SYMBOL COULD REPRESENT AUTISM

The puzzle piece currently is being used to represent autism. It is worn by others to represent the complexity of autism and missing pieces in science. Another symbol that I think would represent autism as well is the symbol of an Onion. Many people have asked me, "Why? What does an Onion have to do with autism?" I tell them, "An Onion has many layers and sometimes the deepest layer could represent speech that is stuck all the way inside. You have to peel away and therapeutically clean each layer that you are working with to get all the way inside." Does that make sense to all of you who are reading this?

In addition, an onion is a very sensitive product. The mere sight of it can make a person's eyes start watering. Autism has its sensitivities. Many people, including doctors, will look at an autistic person and you may notice that they start to squint their eyes, as it is a very sensitive disorder. An onion has to be taken care of very delicately, as a person with autism needs great care and accommodations.

35

Autism and Taking Things Literally

I am sure that many of you reading this book have heard before that people with autism disorders take things very literally. I am here to tell you that it can be very true. People with autism as I mentioned before think very visually, as in what is called "thinking in pictures" from Temple Grandin. Therefore, when someone says something, the person with autism might very well take it out of its supposed context because of the visual thinking and because it can be hard to read body expressions that would otherwise make it understood. For the autistic person, this can make something said as a joke be thought of as literal, and the autistic person may say something they have heard before and use it in a different context. Therefore, in the neurotypical person's mind, being heard very different than the autistic person intended. It can be a cycle of confusion for every word and person.

Let me tell you that there are times when taking things literally can be a very good thing, when taking things literally can be not so good, and when taking things literally can actually turn out to be really funny and cause laughter amongst many people. I have had all three, as most people with autism will encounter plenty of times. At the end, I will mention how to bring along tools that will help autistics start to understand jokes and facial expressions.

When it is a very good thing:

Some people will hear someone making a threat and think it is a joke, not want to get involved, or just plain not think about it all very much. However, an autistic person (many) will immediately but yet not too noticeably (*which is a good thing because autistic people do have issues and most people will just think nothing of the person moving away to another area or to people*) go and tell a person what they heard or saw. These things can help stop things like attacks, shootings and other terrible accidents that might otherwise have been unsolved or stopped beforehand.

When it is a not so good thing:

This was not me, this first example but a kid with Asperger's Syndrome. His teacher was getting frustrated and said, "I'm losing my patience." The kid then seriously said, "Oh, that's no problem. I'll help you find it." Another person said, (*different kid, different scenario*), "Oh, I didn't know you were a doctor." The thing said to the boy was, "He is starting to lose his patience."

Now back to me:

In band, these kids standing next to me had said (*with a smile and laugh that I didn't process because of the autism*), "Yeah, let's just get a gun and shoot the platform down and watch him fall like in that movie." I was crying after school, knocked on the teacher's door and told him. He said to me, "Oh, they were joking honey. I heard it too and saw them."

Also, another thing is when this guy who I cared about a lot who was my teacher and helper said, "I almost had a heart attack last night," I freaked out. I was explained once again that, "It is a joke or sometimes a figure of speech."

There is a commercial on television talking about "If you don't like our tires, feel free to bring them back. Thank you!" In the commercial, this woman throws the tire through the business' glass window. A person with autism may see that and believe it or not, (*this is how literal things can get*) the person, if they were able to drive and didn't like that company's tire, (*one of the tires*) might do what the commercial did because that is what the commercial said to do and what happened. (*This could go for the autistic person who is along with the neurotypical person who wants to exchange the tire for their car*).

This is also a kind of thing where I did not understand it would hurt someone's feelings, because it was just a statement and view. I said to my neighbor when I was younger, "Wow, you have yellow teeth!" in front of other people. This was that literal part of autism and the not so good part of that symptom.

When it is a funny thing:

I had taken a situation literally, really more a phrase, and I did not know it could be taken another way. I was in my ESAP class at lunchtime and said it as a question. The woman teacher, she started choking on what she was drinking and the others were laughing. I didn't understand this; however, it was amusing for the others. They even went down to the psychologist's office and had him come down. They said to me, "Tell Dr.___" what you just said to us." Therefore, I did, and he started laughing as well as everyone else. I was explained a lot later what it meant, and boy was it pretty funny. Actually, I still do not get why it was that funny, but it was. So, whatever ☺.

This next story sort of has to do with taking things literally, and I just had to add it to this book as many of the people said in my family laughed for a long time and said, "Yep… she's autistic and that was a great autistic moment!" (*This was meant in a good way of course*) Here is what happened:

My brother bought a watch at the Mall of America that he had been saving up for after hard-earned extra put aside pay, that he deserved and ultimately wanted. It was a watch that cost him a couple thousand dollars (*and was on sale and a deal at the same time*) to go with his work clothes so he could feel special. I had bought a watch set that was interchangeable (*I mentioned in another chapter*) and had basically every color and every watchband to go with any outfit (*fancy or casual*). I came up to my brother and said, "You bought a couple thousand dollar watch that you can only wear with your work clothes but I can turn this watch into any one I want for any outfit I want and I got it all for only $8.99!" Well, it is literal and the truth…

As you can see, this "taking things literally symptom in autism can have three areas: When it's good, when it's not so good and when it's funny.

As I said in the beginning, I would add tools for helping people with autism be able to better understand jokes, as well as facial expressions.

1. Therapists have asked me from other states for their "Autism and Asperger Syndrome support groups" what I might suggest to help teach their clients to help understand jokes. I tell them that there are a few older series of shows that I recommend to help and they are Bewitched, Dick Van Dike show and Hogan's Heroes. Bewitched and Dick Van Dike are the ones I recommend the most. I and another group have really found it helpful for being able to actually understand jokes finally!
2. For facial expressions, there can be facial expression cards in which you can do skits with. There is also teachings now and small clips from a new TV series, "Lie To Me". It is an interesting twist on how to tell what expressions/ body movements express what emotion, etc. I have seen it a few times, and since many of the expressions and movements don't have to do with using direct eye contact, it has been interesting. Any autistic person can utilize it more since direct eye contact can be very uncomfortable. It can take time to get used to but

can help by taking tips and clean clips from certain shows. You can even use screen shots and then explain yourself to your group. Just as an add, I want you all to remember that our facial expressions may not match how we are feeling, so this can help as well.

3. One other thing, if you put on closed captioning, you can see that it will say things like, "Nervous cough." This can help bring along awareness to different sounds, etc. and what they mean socially.

36

AUTISM AND QUIET THINKING

Many people with autism tend to think a lot. When I was in a support group, we tended to think a lot about how to stop problems in the world and came up with childlike quality, simple ideas. However, of course, some of the ideas are simplistic but may not tie into capitalism (*how it was maybe designed*). It would not be surprising to find a person with autism constantly thinking about things.

In our group, both I and us came up with simplistic ideas and the rest of the people in my group thought it would be a good thing to share in the book. It can get others thinking about it as well and to inform others that many people on the spectrum think and think and think!

You might ask what goes through my mind so I will show you this time… just a little bit of what I/we think about.

1. I do not understand why not all people just burst with excitement when they are happy. Why do the majority of people think it is strange at times?

2. Just to stop in time, close your eyes, and to feel the spirit and the air surrounding you, holding you always.

3. We could see the children running in the wind and follow their simplicity in joy.

4. Money is material… it is printed. We constantly talk as a country about how we are in great debt, and a childlike thought would say, "Why don't we just print more?" -- Of course, they would have to mark it with a special ink so people wouldn't just print fake money. --
5. If we stopped and realized that ending poverty is within our reach, within days, if everyone in the world just gave $1.00 we could help take away much of the poverty. What about $2.00? It is not much to give. For a small thing given can amount to a large thing in return. Why hasn't it happened yet?
6. Why do people in a marriage have to think twice about trust if they already know the deep roots of the person they're with? Why does love have different uses?
7. People rely on words (yes words are important). But, sometimes there are no words because what you feel exists in the realm. When you listen, you can actually hear the impossible, the whisper. It says absolutely nothing and absolutely everything at the same time.

Those are just a few things that I/we in the group came up with. The other things that we tend to think about, some are more technical and some are even more simplistic. I am sure that some of them are because of the autism and our not understanding social things or government works, but if you really think about them, don't they make some sense? Many times when you see a child or adult with autism sitting there with their eyes closed or staring off, we are thinking and sometimes we are just listening to Mother Nature. Therefore, our mind can be completely stateless so there is a wonderful peace. Sometimes people on the spectrum think too much and the thoughts and scenarios make you feel like you are going crazy. A talented and well-known musician was seen rocking and stimming. They realized later on it was because he had too much music ideas and sonatas going through his brain that it was clouding him, making him stressed. Autism does have its struggles but there is such a simplistic and peaceful side to it as well;

the side that I wouldn't give up. Many people ask me the question, if there was a cure for autism would you take it? I add just a touch more coming up later on in the book in the chapter about the +/- of different autism diagnoses.

37

AUTISM AND VOICE PITCH

People with autism spectrum disorders can talk either too soft or way too loud. This is really actually a common problem, or rather, a hurdle to try and overcome. Sometimes it can seem impossible.

Why do autistic people have this hurdle with their voices, and more importantly, do they know what pitch or volume they are using? I am going to answer those questions for you below.

1. It is a cause and effect problem
2. Vocal cord development
3. Sensory issue

An autistic person might not actually hear themselves well, since it is coming from the inside of their body, therefore having their pitch be loud. On the other hand, they may have soft pitch because it can be really loud to the autistic person's sensory since the waves of their voice come out but bounce back up to their ears. It is once again, an area with a very scientific explanation and does have everything to do with autism.

Do they know what pitch or volume they are using?

In a sense no, because we will be having to be reminded or told what to do, but in a sense yes because we are told so much and when we try to use another pitch it can become the complete opposite. So, we can understand in that people are telling us we are using one extreme to the other.

I used to get angry and frustrated because I was told to do one thing and then another. It was a cycle that just kept happening and happening. I actually tried to give up and sometimes I just do because it is not my intent. People, (at least through my book) I hope will be more educated about this. I hope you will understand that the person with autism is neither trying to be mean or obnoxious, nor trying to ignore or become distanced in your conversation. It also does not mean that they are scared of you or shy either.

I really hope this has helped and once again, I hope you can pass along this information you learned from my book to others.

38

Autism and Physical Features

In other clinical DSM disorders, there are sometimes some pretty visible/noticeable physical features. In autism, there are some, but sometimes you need to be a psychologist or need to look very closely to notice. Some of the physical features can be quite easy to identify.

I am going to give you a list below of some physical features of autism that you may look for in your child or that may actually turn up in tests.

They are as follows:

1. Irregular EEG – slower blood flow in parts of the human brain which means slower processing, comprehension issues, etc. (*not life threatening*)
2. Larger pupils of the eyes – This is also one of the reasons for our light sensitivity (*especially after birth*)
3. Different affects inside the ear – This is also one of the reasons for our hearing sensitivity. It is common in autism to get constant ear infections and have issues with eustachian tube draining. I want to mention here that there is a special product and a special type of gum called xylitol. When chewed,

it travels from the excess saliva, up through the throat into the secretions of the ear that are affected and terminates the infection. You may be wondering what types of gum have this and not very many have them, so you have to purchase it through a special infection fighting company. However, they have added to some stores the gum called "Epic." Ask the pharmacists where to get them if you cannot find them in the drugstore. There are rinses and dental things to help as well. It helps so many things and is natural! Go to homesteadmarket.com and go under xylitol products. It actually is not too expensive. In the meantime, if you can locate "Epic" gum in your drug stores, that is a good start.

4. Very skinny body type, sometimes it is the opposite: This is because of low muscle tone/leaky gut that can cause the weight part of the human body (*the holding of proteins, etc.*) to leak into wrong parts or not hold (*weakened cell membranes causing oxidants during the fight and pushing through, resulting in colonizing other areas of the human body*). Therefore, it can cause thinness but for many who have leaky gut, it leaks into the wrong parts creating heavier weight.

5. Larger head (*this is usually more in the birth/after birth stage and can finally grow into the body*)

6. Red ears (*as well as larger*)

7. Fading teeth enamel – tooth decay

8. In quite a few people with autism pelvic bone slipping (*obviously an inside physical feature/bone and brain affect*)

9. Sometimes there is a specific golden ring around the pupil that looks kind of like a sunburst. This can indicate Wilson's disease, but they have found it can be a symptom now of just autism since the two syndromes are similar in nature; they can coincide.

I hope that these physical symptoms can help you notice signs that might indicate autism and be a backup for a diagnosis for your child. They are the "physical" signs of autism, as there are physical signs unique to each different disorder. (*Unique birth as I like to refer to rather than disorder at times.*)

39

THE +/- OF DIFFERENT AUTISM DIAGNOSES

There is profound autism all the way to Asperger's syndrome.

I want to caution people that in some movies the actor will say, "He is high functioning autistic." Most people who have children who are diagnosed as high functioning autistic don't see much or sometimes any resemblance between the person referred to that way in the movie to their own children. Therefore, many people get confused and some even start to question their child's or family member's diagnosis. The reason this is so is because in each level (*classic, high functioning, Asperger's, etc.*) there is a level of its own. In some of the movies, they should have made clearer as to say, "He/She is high functioning for having been diagnosed in life as low functioning autistic." However, many of the movies have failed to express it in that way and even doctors, as they know what they mean but sometimes forget they need to express their guidelines for the diagnosis in more detail for the non psychologist.

I will also like to say that in each level diagnosed, there are problems of their own. Some people will say, "Well, since your child is high functioning or if your child has Asperger's syndrome, then they are aren't affected in as many negative ways so it should be easier for them." This is true in some ways, but each level brings problems of its own.

For Ex: (*Please remember that each level has its own level – Often time Asperger's Syndrome is referred to as the highest level and least affected as the mild version, but some people with Asperger's are not compared to other people with Asperger's syndrome who you see getting married, etc.*)

1. Asperger's syndrome – This can bring along the knowledge of being different which can cause great pain and can also bring about really high expectations that people don't realize can be extremely difficult (*some think that people with Asperger's should be able to learn pretty easily finances, other living situations, etc. that can be hard as it is still a part of the autism spectrum*). Some people will say that the person with it shouldn't and couldn't have those vestibular or sensory problems. Or, they say the child should at least be able to hide it. It can also bring about the person with Asperger's getting literally exhausted in trying to pretend to be normal. It can also lead to instances where Asperger's for some can become almost invisible, so they have to deal with explaining they do have a disability and not full-blown autism. This in some cases creates humiliation, uncertainty and problems.

2. All additional levels of autism – profound autism all the way through moderate autism can bring along problems amid severe meltdowns, problems with fear of hurting one's self during the meltdown, severe vestibular and sensory issues, severe difficulty with apraxia, a variety of other problems such as a condition that causes cravings of dirt/soil, etc. Some people with a diagnosis of high functioning autism or alternatively the Asperger's syndrome (*remember again, different functioning levels have again their own levels*) may sometimes say, "I wish I was lower functioning so I would not have to understand." Then there are those that have higher functioning autism and who say, "I actually wish I was lower functioning so I would be treated more with understanding. I also wish this was so, so that I would not understand that I'm different or understand some things that really stress me out."

I want to say that these issues can cause great depression and great anxiety in people, especially school-aged kids with it. Some people with autism will say things or act out (*behaviors*) that would indicate they are overly frustrated and feel like they can't take things anymore. There needs to be methods in place. There are different sides to when a person with an ASD may say or do things that for a neurotypical population would require emergency services. Sometimes the person is stressed and they take what they have heard (*such as self-threatening remarks, etc.*) and expose them to themselves and those around them at home and in school. There are other times where the person is very serious, and with facial expressions and not knowing (*including bullying*) what is going on, you have to take it seriously. A great resource would be the people who are around your child in school and during the day to see what is triggering these threats. There are also cases where the person with an ASD or related disorder is just punished when acting out because of being overstimulated, schedule changes or for another reason. For what is seen as very inappropriate behavior that warrants punishment in a neurotypical is different when it comes to a person with ASD. This is because it's considered normal for their body to react that way. It makes me cringe to see a person with autism being punished or even security guards getting in the way, as that only makes it worse. They should have a professional work through it with the person. Most the time when acting out happens, it is a true cry for help and the screaming is translated into, "Help me…" I had security guards that did not understand my behaviors but thank goodness the professionals were there to tell them what was going on and to work through it with me on my level. I think every school needs a room just for kids with mental health issues and disabilities like autism. This room would include therapists when they need be there, swings, other therapeutic and calming techniques, and basically be a safe room to solve issues the correct way. I had been in a classroom (*mostly students without disabilities*) and this kid was acting out due to changes in schedule and a comment made to him. The teacher was going to discipline him and he started getting worse. She was getting a bit frustrated, and I offered to take over (*since I knew what was going on*). I took him outside the classroom and brought along a calm and a nonchalant mood. I had also brought along some things that I thought could help if the situation got any worse. The transition getting him out of the room was hard, as he was visibly getting more tense from the tense feeling he was picking up on. It only took about five minutes and he was then fine. The teacher was interested and so were the others who had dealt with his emotional issues. I watched them the next times and told them what might be better to do during the issues. Sometimes you have to be a little creative and I also believe in aids/professionals going to training as well. I hope to bring more training or words of knowledge to other conferences as I have spoken at one before where there were a nice variety of people. All schools need to come up with an individual "visual" box of some choices the child with autism can rely on when the meltdown is coming on or is starting. It gives them back more control.

People with high functioning autism can have almost all the issues of a person with low functioning autism and therefore can almost truly be considered medium to high functioning autistic. The different levels, since they each have their own levels, are what make autism so very complicated in so many ways.

At home, there have been times where I am supposed to be filling out a check or just transferring money and I go into a complete meltdown. My parents then again realize that I still do have high functioning autism (*remember once again each level has its own levels*). Even other people I know who have Asperger's will storm and trash things around them, screaming because it is still too hard to understand. They can just start to get overwhelmed.

My parents realize that they have to stop explaining because I really honestly do not understand it and am too disturbed at the moment to comprehend any word they are saying. Also, sometimes these things that can come along you need to understand may take a long time (*years later*). You should also know that it can be much better coming from a rehab person or a psychologist who will take it real slow and nonchalant. They seem to know when is the time, and when is not.

I want you all reading this to know that each level brings its own issues, so when you hear a person say, "She/He should have less issues because of their diagnosis level," etc. you should then correct them. Let them know that in each diagnosis level there are not only different levels in each level, but also that each functioning level brings along its own unique issues and struggles. You could even recommend them to read this chapter in this book I guess also. Let them know this is a person who has autism trying to explain this and that others I have spoken with on all different levels of the spectrum agree with me.

I would also like to add as part of an ending note to this chapter, to know that the word autism included in the diagnosis many of times receives more much needed therapies and accommodations rather than a diagnosis of Asperger's syndrome. However, with the different levels as well as brain differences, Asperger's syndrome will hopefully start to gain the same rights and as much help as a label that contains the word "Autism." It is a part of the autism spectrum.

In the upcoming DSM-IV-TR versions, the psychologists are trying to decide to put the disorders in a list, putting "Autism" as the one diagnosis and in gray putting the different levels. This is so that each level will receive the same amount of services and be looked at both differently but the same.

40

Autism and Being Behind

I try to explain to parents that many times the child with autism is behind. At the age of twenty, the person with autism (*depending on the level in each level*) could be more at the age of 14 and just starting to understand things more and become more independent. However, they would still need lots of dependence and support. This also goes for "the terrible twos" stage. At the age of two, many parents have noticed that their higher functioning kids with autism did not have that stage up until roughly age 7. Similarly, parents have talked to me about their child being very rigid in their thinking at around age 9. They are experiencing the autonomy stage, or the stage that neurotypical children would go through at around 4 to 5. This stage is basically all about independence that causes conflict, or an independent identity and power struggle. Therefore, you can see that the age for the autistic person (*development stage wise*) is slower to appear, so milestones can be late. Unfortunately, with the autonomy stage, many have to deal with it at a more difficult time when the child is bigger and stronger in defenses.

I am going to talk about autism and bathing. I had a parent email me that they were concerned because their six-year-old child with Asperger's Syndrome was not bathing himself alone yet. There are so many people going through the same thing. I, personally, did not bathe by myself at the typical child milestone of doing so. One is because of taking things literally and the other is because of the autism factoring into it.

As I said before, a lot of things are behind by at least 5 years for many people with autism. I was finally just starting to bathe myself alone half way through junior high school. There are not too many people probably who would write half the stuff I have in this book because of friends and family members who live away. However, I've decided that I want to help people deal with autism and understand it better. Secondly, I believe that if these people really love me for who I am that this will have no effect on whether they want to associate with me. For the literal part, I know that many families in other countries have a big bath (*almost like a Jacuzzi*) that they as a family bathe together in, so I did not understand that it was different or unusual to the American culture. I think all cultures should be respected!

I am now going to mention autism and toilet training. Some kids with autism have a real problem with training as it does not make sense, so you need to spice up the training in other words. For one thing, as you have already heard, autistic people are very visual. You can purchase one of those dolls that you feed water to and then it pees. Once your child is watching the doll "pee" in the toilet, then give something (*that interests your child a lot*) to that doll every time it does so. You need to do this several times (*and spread out as well throughout the day as you work up to it*) and you need to make sure that the rewards are intriguing for him/her. Also, have him/her sit at least even once on the toilet (*even if they do not go, give them a small reward – not the most enticing, but something he/she likes*). If he/she goes in their pants, which for a while they will do I am sure, take their pants (*and have them follow you*) and put them on the top of the potty. Then squeeze them out into the toilet so that he/she visually connects seeing "what he/she goes" as belonging to the toilet. Consistency and visual presentations are key in helping this process along when teaching your autistic child potty training as it can be one of the most difficult hygienic struggles with the disorder.

I will now talk about general knowledge. Up until just about a year ago, I really did not understand many things that typical people would understand such as certain words, phrases and what they meant, jokes, etc. I also did not understand some social graces like smiling. Also, I did not really understand that I always had to answer someone back. I figured that they told me or asked if I would do something, so in my mind, it was a statement and not a question. I still have difficulty with that today.

Similarly, this is yet another thing that most people probably wouldn't put into a book. I want to help answer possible questions that run through parents' and families' minds that are not typically talked about. That is what I hope my book is bringing as well.

I take things very literally and since I am a Christian, I have heard the story of Adam and Eve several times. And, as a part of it, Adam and Eve in the beginning didn't really wear anything except a cloth. So, in my mind, I did not understand until a few years ago why it was such a big deal if you didn't want to wear clothes. Once I realized this countless years later, I felt upset but then again I did not because it was a part of autism and that is just the way it was. Plus, I guess as a little bit of humor, I have an excuse for why I didn't understand – it was in the Bible!

I want to let you know, that up until not too long ago, I heard that "things" had feelings. I would hear on a television show or in a book or movie, "A ___ does not have a home or feel at home until it is cared for." So, whenever I would be doing something I would be sympathetic, even towards cereal and paper given that they were "things" and "things" have feelings. I did not understand for so long that those "things" were things that breathed and felt, not just "general things." I do have to say that since some people with autism have a childlike mind, they are curious and think of how it would feel to be "something," I still have these thoughts and act sympathetic to almost everything. Then I remember again that these "things" are "breathing things." Still I have that so wonderous mind of just wanting to know scientifically everything, so that interferes as well. Therefore, if you see your autistic person you love doing these things, it is because they are wondering and thinking hard or misunderstanding. On the other hand, they could be using imagination for just regular things like teddy bears, which is really good since imagination is usually not very good with autism.

I want to add here about seeing or hearing something and taking it to heart when you are autistic. You are not really taking it to the "logic" part of the brain because the autism takes over with innocence and the chronological age differences played in to it. I saw in those commercials for Carnival Cruise Ships all these things like beaches, amusement parks, bowling alleys, circuses, etc., I saw it for what it was and thought at 22 years old that all that stuff fit in that ship. I looked confused but I figured that if it showed all that stuff and it talked about going on that boat to be able to do all that, that all those things must somehow all fit in the boat. Obviously as I now know after many confusing looks, stares, and me finally asking how they fit all that into the ship, that I explained the "logic" of the commercial. Yes, I did feel embarrassed again, but then really not because what I see is what my brain as an autistic person sees and accepts.

Many people with Autism will put several things together that people would think to themselves, "Did they dress themselves in the dark?" I used to put jeans or pants and then have a shirt with a kind of casual type of dress (*you would have to see it to know what I mean*). I did not understand they didn't go together. There are now really great things to do to help people with color blindness. You can buy animal characters and place them on hangers. The giraffes would be put on hangers so that the person knows those things can go together. It really can help. I am now getting a little better at that part too. I would like to mention here that part of this has to do with sensory and what feels good together at that time. Sometimes a person with autism can love a pair of pants or a shirt one day, and the next, scream and throw it across the room. It has to do with the senses and how they are regulating at that specific time.

In conjunction with that, there are many people on the autism spectrum that have trouble with hygiene. I have a few tips for that as well.

I am 22 right now, and I still have problems with this sometimes. It sometimes is a result really of a scheduling problem as well as getting distracted. Also, when a person has Asperger's syndrome or another autism spectrum disorder, in many cases we don't really feel our bodies much. Therefore, we do not feel like we need to wash up because we don't feel what others do when they say, "How in the world could you not take a bath or brush your teeth or face for even two days?"

I really recommend the mix and match watch set from collectionsetc. com because it has almost every color in bands and loop decoration. My mom said to me finally, "You're not wearing the same shirt and pants over and over for four days in a row… you are actually changing clothes at least every two days now." This is because I started to get excited (*as well as its getting older chronologically with the autism*) and maturing just later, so I can mix and match a watch and wear a pair of clothes that would look good with it! It is sort of a motivator thing as most things are to people with an autism spectrum disorder. As I had mentioned in the chapter about tactile defensiveness, teeth brushing and reinforcing through the senses by using a vibrating toothbrush or vibrating bristles through z-vibe and tips helps. For being behind, that is very difficult, but another suggestion I make is a toothpaste that looks appealing to the visual field of the senses. This is usually very captivating and helps a lot. The brands that can create an actual wanting to try hygiene in the area of brushing is Kid's Crest and Wild Expression Toothpaste. These specific toothpastes have flecks of mint that look glittery and are more like a gel, have a flavor without it being harmful to teeth, and do the same amount of protection as the other toothpastes. It is a lot feeling the body and reinforcing the senses that can help when dealing with the ordeal of being behind in terms of hygiene.

For brushing the hair, most the time this is because our scalps are very sensitive to pulls. Therefore, getting a conditioner that helps with tangles along with using the correct temperature of water is important. There is also the times where you need to fix your hair and you have already had a bath that day and do not need another one. There is stuff you can get to spray in the hair that makes it easier to get through.

Also, I have mentioned that it may be hard to get your child in the bathtub and then hard to get them out because of the feelings. This includes the wet feeling of the scalp and hair that can feel good after the initial transition issue. Some parents are saying, "My child will not go into the bathtub and they will not let their hair be brushed except for once or twice a week, and even then it is a struggle..." I have advice for that. As I mentioned the transition issue, I also want to mention that if you continually fight the struggle with praise and rewards every day, then the person with autism will get used to bathing and hygiene every day. It will start to become part of their automatic routine. They will actually start wanting to take baths and use hygiene techniques that neurotypicals would use every day without even thinking twice. It is because it is a need for sameness, and then it becomes engrained into the schedule. Sometimes aromatherapy, definitely not too strong (*because of the hypersensitivity in autism*) can help. There are specific scents that stimulate parts of the brain to help with excitement and have been proven to help even with autistic people's brains. They are using this as a new therapy now as well for many autistic people. It can work wonders not only for excitement and encouragement part but also for sleep and relaxation that is sometimes used with color therapy.

Knowledge:

In a way this is good and in a way not always so good. A person with autism may not understand certain cuss words or that a certain phrase or topic is considered not good. They might laugh when others are laughing at something that they think they know about but should not. They can then go off and repeat what they heard. That is the problem with that, but at the same time, at least your child with autism will learn about it appropriately at the age a person *should* learn about it and then respectfully acknowledge it. You can almost look at it as a better and pure thing. However, when in conversations, sometimes people may look at your child or young adult with autism and not be aware of why they do not understand it. It can become a problem in that manner. However, this is what comes along with autism a lot of the times.

I am now going to talk about walking. First off, for people with autism, it is not unusual for them to have balance problems from vestibular problems and have low tone in their muscles. This usually comes into play when they are not reaching their milestones. I was a late walker. I was walking at about 14 ½ months and in the video our family had taken, I was tripping or falling even around that time. It was like beginning. That is not uncommon but inconvenient at times.

Along with late walking, many people with autism have an abnormal gate. Sometimes they will slug their feet along and it makes noises on the floors. It can also be a bent over type of walk as well, as the feet can almost go inward at times. Sometimes when a person with autism runs, and in Asperger's syndrome, it can almost resemble a frog type of run. That might sound like a harsh way of putting it, but I don't mean literally like a frog. You would have to see it to know what I mean (*and I think that some of you reading this book will understand what I mean by that*). I always was made fun of when I walked and ran and sometimes got into trouble because they thought I was scuffing the floors on purpose. Once again, a great psychologist explained this to help everyone out.

One of my psychologists wrote agendas for our sessions. I would write a poem about why it was good to remember certain things or draw pictures for him during the therapy.

I am not going to say too much about this, but the phallic stage can come on late as well, very late in fact. This can last a short time or can be utilized by the child and even baby with autism, unknowingly thinking it is tactile input.

In the next chapter, "Can he/she live on their own?" I bring to light some tips for that, as autism and being behind ties into it as well.

41

CAN HE/SHE LIVE ON THEIR OWN?

First off, I mentioned slightly this answer in the chapter about autism and love. I also will touch on it as well in the chapter, "Autism and change.' Here, however, I will explain to parents and or caregivers how to deal with this issue, especially if their child is higher functioning.

A parent has instant messaged me and asked, "How can I get my son to start thinking of moving out on his own? I don't want him to think that he can't just because he has autism."

My answer to her question was these things below:

1. This is a hard one. For many people who are higher functioning, it is really about six to ten years later (*or more*) if it is even possible.
2. If possible, have him help you with the cooking, laundry, etc. and give him some time to take care of himself for a couple hours with you somewhat nearby to see how things go.
3. Get in touch with a life skills person.

The same parent said to me, "Maybe I am just not being firm enough with him." My answer to her comment on that was this:

No, I would say it is the autism aspect. For a lot of people as I said before, it takes many more years even with really high functioning autism.

In addition, it has to do with change as well. For me, where I live is where I want to live forever. It would make me sick to live in a different environment. That is the problem and for families they fear that because they know about autism and not handling change well. That is why I suggest respite, families understanding to stay in the house and not sell it, angels' services, etc. This is all so the person with autism can stay in their environment. I think they should put plans together in the government for people with autism disorders who the transition would be traumatic for if the caregivers pass. The government could put these programs above in place as well as let the rest of the family who would usually get the money split from the selling of the house, money from the government because of the sensitivity of this disorder "autism." Do you who are reading this agree? This is one of the most common concerns to parents of autistic children, but in this book that is why I am providing the information and also asking you to write to your chapters of autism and your council people in your area where you all live. Hopefully you can write to the President of the United States or wherever you come from! It does not have to be as scary as it does seem, as you just need to read the resources and get in touch with these places. They go slow and they really care. And hopefully, like I said, new programs that I want to happen can happen through this book by me and all of you. It takes one person to start making a difference and that can be you! Sometimes group homes, the person will transition to (*you should have your autistic child be near the area so they get used to it and the people there to help*). But, for some autistic people, it will prove to be too much pressure and result in psychological trauma depending on that area in the autism and how it affects them. I think in some cases, the autism research and walks should go towards creating programs for this and as well as all the chapters for autism in each area of every state and county. It could help bring on a familiar environment by taking everything familiar and making a big ranch for autistic people who have trouble with transition. Take the actual objects in their home and make it their own space, each part of the ranch or place so that it is really still "their house." It is only moved and set up by taking pictures and doing a layout. I think it is well worth it for this. We can do this as a community and as families. I do not want those of you who have decided with peace on having your child with autism go into a group home to now feel really worried, as your child with autism may take to the transition just fine. I am just giving many alternatives and ideas to the rest of the public. They are for those whose autistic children get physically ill and traumatized about change and cannot overcome it even after a long time. For other disorders besides autism, it all depends as well, and we want as a people to have the best at heart for each individual. We just need to use our minds to come up with creative ways to help each person differently!

Now, about higher functioning kids moving out on their own or with assistance...

I would not push too hard as it can break a foundation for possible moving out, as it can panic the person with autism or make them upset. Then, their chances go down instead of staying steady or going up. It just has to be slow and it is better if the person with autism brings it up. When that happens, just talk positively and let them do more of the talking.

There can be great potential when there is a service dog in place as well trained by great professionals.

There are other directions a person can take and utilize as well when thinking of having their child with autism live on their own.

Brothers with mental disabilities had their neighbors and state checking on them so they could all have a house even with the limited ability they had combined.

I will give you other living examples below.

1. My parents said that they could buy my neighbor's house maybe and they could go in that one while I stay living in the original house. This gives the person with autism their wanted independence but also helps with the needs that come along with it.
2. Lots of Native American tribes use this type of idea as stated above because they are very family oriented. There was an example of this on the show, Extreme Makeover Home Edition. The houses are connected, yet separate.

On my website, I have a slideshow and it shows the many different alternatives and directions this topic can go in along with pictures to help explain as well. I would highly suggest viewing the slides as well as taking my suggestions from this chapter.

42

AUTISM AND CHANGE

I would like to mention that the word "change" and autism, if anyone knows even little about autism, knows this can be a very major area. This most of the time is the case no matter how high the functioning level. In this chapter, I will be talking about several types of changes and as a bonus, talk about autism and bringing a new family member into the household!

This chapter could extend on for several upon several pages. I know that earlier in the teaching section I brought up seating changes and talked about how I would expand on that in this chapter for the college segment. I also talked a little earlier on about specific noises that many (*more often children*) make if a change is made to a "line" of something, but I will expand on that further as well. I talked about environmental changes, such as moving, but also have another side to mention as well. In this area, it may seem somewhat surprising, but I am going to specifically bring up two other areas of change that are not talked about as much. This reason is that they may seem like separate issues in a way to the subject of autism, but as you will see, are not. They are the transition to new movies, shows, toys, etc. There is also anxiety in change to visiting new stores or even websites as well. Yes, even visiting a new website can be stressful because it is a change.

Seating changes (*the story in college*):

Several students with autism have (*or should have*) priority seating, which is seating that is best suited for the student and usually occurs where it is either quietest, least distracting, closest to the front and where the teacher can refocus attention. I did mention in the college section of the teaching chapters that I tried some college. I had a class and the teacher knew about autism. He was well aware of the problems that could occur if I were to walk into the class and my seat was accompanied by another person. He did a smart thing and put all of his materials he needed for class on the seat and the desk part of my table so that everyone else there would automatically choose another seat since you do not move your teacher's things. However, one day, he forgot to do that. It was somewhere near more of the beginning of the semester yet leaning more towards the halfway mark of it. A student was transferring from a different class to our psychology class. She had took my seat because it was open. The seat across from her was open as well. I walked in and my teacher had not noticed quite yet what was occurring and did not realize that I wasn't there yet. I started to walk into the room and of course, when I saw this, I started to get very upset inside to say the least because of the autistic brain and its various chemical balances that are different (*as well as flow, etc.*). I went up and said, "I have priority seating." (*Well, at first, I stood there and the table was staring at me but at least they knew me, the three out of the four of them.*) The new student did not really understand so I made up that I had trouble seeing (*at this point my teacher had realized what was going on and knew he had to help with his knowledge and charm*). The girl was saying that she was understanding but said the seat across is the same distance away, just on the other side of the table. I thought I was going to have a heart attack. I said, "I can't. I have..." Then my teacher interrupted and said, "Oh, there are some other circumstances and please just move over to the other side please." I guess she understood because all of a sudden she moved, got really excited and smiled at my teacher. The other people in the room already knew that I had autism because I guess when the class was learning about it, it was very obvious and then later on I gave a presentation with his help and the 3 other people at my table! And I believe there was a similar case to mine in the class as well but they didn't quite get that as soon. There were a few people who just didn't connect with it, but the others had figured it out (*I had a disability*) after the first couple of weeks. So, this is a very important story that I thought you would be interested in knowing about as well as it being another lesson and tip for teachers:

1. Teachers, put your stuff on the chair and desk so that your autistic student doesn't get his/her chair accidentally taken.
2. You can teach the autistic student to say they have priority seating which most students will understand at that level and age.

Now, I will extend more on the topic on change in "lines." A lot of people are baffled about what reactions can occur from the autistic person's body when something is taken out of a formation or "line," and as will be explained a little later, even if something is stopped. However, I will give you a few examples that I hope will help explain this symptom that is quite common, especially in younger children. This is a visual field to brain reaction.

Examples to help explain these automatic reactions (*or otherwise similar to automatic-type reactions*):

1. For a person who has seizures, the visual field takes in light and brings it to the brain, creating a reaction; quick movements of the person's body that are uncontrollable.

In autism disorders, the visual field sees movement following a line that is somehow connected, sending the signal to the brain causing their reaction to become rough up and down sounds of the voice impression. They are automatic-type reactions controlled by the brain. It can happen from acute fear.

I hope that helps you understand better why this happens and what is really going on.

I am now going to be talking about the side of change in autism regarding introduction of new movies and toys. You may find that your child with autism becomes fidgety and upset even by having to go to a movie (*they have never seen it before, therefore they don't know what to expect*) and this is the other side of autism and change issues. This can even be with toys. Some people with autism don't like to be surprised and want to see things and pick things out so they know what they are getting (*like at holidays, etc.*). This creates repetition that the brain can remember and not have to sink into newly.

This even goes for visiting websites; yes, even visiting websites. There is a person who asked me if it was just her or if my website at first bothered other people and I told her, "I was upset creating it… because I didn't know what to expect." It can be simple things like that, but they are not simple in autism.

I tell parents that for their child, they can gradually bring things into their normal environment (*while the autistic person is happy and also when he/she is not noticing it happening like as while distracted*) that resemble people or actions of a new TV show or movie (*or even scenario or toy that is new to the autistic person*). This will help to lessen the anxiety and produce happiness in the end.

I could talk about routine forever but will just mention a touch of it here.

1. There are instances where an autistic person may have a television show that they have to see and if there is a diversion, it becomes upsetting. It helps to have a DVR to record so the person can be sure that it will be there. It is not entirely missed and there is also the other opportunity to get a small portable television so the person with autism can take it along with them if going somewhere in the car to an appointment.
2. There is also the need for routine when going from home to school or from home to somewhere else. I would get very upset

if my parents who were driving me would take a different route or turn, even if it was just one turn and the rest was the same.

3. For the autistic person, their bathroom is almost like a sanctuary and safe place for them. Many times they go there to organize things and relax on the cool floor, etc. It is important to keep this area the very same, because even just moving a toothbrush can be a catastrophe in some autistic people's minds.

4. We were at a pizza place and my parents wanted to sit at a different table and they just forgot for a minute that I do not like change. I sat at my regular seat and they realized and actually laughed. They came to sit down and I told them they weren't sitting in the right seats so they moved to where they sat before. This can go on to be even more with OCD and can lead to having a group or support group having the autistic person worried about other people sitting where other members sat before so wanting those spaces empty. It is so difficult.

5. I mentioned this earlier, but a person with autism can be so upset with change that either they have an accident or they throw up. Many animals will have this happen if their environment changes too much or their bathroom area is moved. I have had my bed changed (*outgrowing*) and I had slept on the floor for about 6 days instead of in it. It freaked me out and I had even thrown up once when more recently a new television and setup was brought into the house. Believe me; I do not deal well with change.

6. I now want to mention that many autistic people will have a store that they go to (*to help mom & dad, for fun, etc.*) and they become attached to that specific location. Therefore, let's say that your child goes to Fry's Market Place on 154th Gary Drive. If you were to go to a different Fry's Market Place, your child with autism could get very upset or very anxious. I have gone to a different Fry's and different Walgreens, etc. It really makes me feel sick because I want "my place" that I am used

to. I don't like to go to the same place if it is in a different location because it is a part of change with the autism factoring into it. It may seem quite strange, but it is an attachment and a need for order and having things look exactly the same such as the same familiar faces and aisles, etc.

For that special part of change I mentioned earlier: Bringing someone new into the home! When a new baby is coming into the home, it is difficult on even neurotypical children because they get confused and sometimes jealous. However, when you are autistic, this can be much more serious. I have advice for children with autism who tend to be over the top (*overly excited*) and also who are confused who get upset.

Here are my tips below to help you and your family transition more peacefully:

1. Many children with autism can be more hyperactive or more aggressive with a baby, not understanding how fragile they are. I would reinforce "gentle" with PECS. You should have someone modeling "gentle" around the baby such as quiet voice, soft rubs on the arm and being careful around the baby's stuff. Then you can help hand-on-hand the modeling to help the child.
2. Some children will seek confusion but yet overindulge in getting a reaction from a small baby (*to the autistic child, the baby may seem more like a creature rather than them at a different age; It is the comprehension in the different levels*). However, in some ways this is good because many children with autism do not have the natural ability to seek out humor or seek out reactions to their actions (*cause and effect*). In order to balance this out, you need to take what I gave advice on in tip 1 and incorporate it into this tip. However, I will still reiterate here and also expand into what else needs to happen for this scenario. I would suggest using "humor PECS" around certain parts of the house and "gentle PECS" around the baby's parts of the house. I like to encourage parents during this to have

alone time with the autistic child where you can model fun and humorous activities, like peek-a-boo or whatever will get a smile on them but not over stimulate them. Then, at the end of the session, do something calming like swinging or rocking and remind him/her of being gentle and soft around the baby. By soft, I mean soft voice, like a whisper, which you would model. Just combine the words baby and gentle/soft together a lot to engrain it into the brain.

3. Many children will want to kiss/hug their little brother or sister (*you can also let them know they are a big brother or sister – depending again on level of understanding*). I suggest that you let them kiss their little hand or above it so they don't accidentally bump the baby's teeny eyes. Just a tip from me, LOL, even though I am not a mother.

I could go on and on for probably 10 more pages on this topic alone, if not more (*as well as the other chapters*), but I hope this has shown light onto why change is so hard (*even when it seems small*) when it is mixed with autism.

43

AUTISM AND OCD

OCD is very common to be associated with autism. Autism is actually a bunch of disorders put together as I have explained before by incorporating the word spectrum. Just to make sure that everyone is clear on this, OCD stands for Obsessive Compulsive Disorder.

There are many sides to OCD. Some things in autism such as tactile stimulation as well as lining things up and even the stimming can co-exist with the OCD. Yet, it can be unique in itself to autism.

Also, many people who have OCD have trouble explaining their OCD even with lots of coaxing and therapy, so for the autistic person, their OCD is going to be far harder to explain and decipher. Hopefully, with the scenarios I am going to show you here of the different sides of OCD (*that took me lots of time and help to get out and still can't explain all of it*), it will help you understand some of the autism and OCD co-existing behaviors that are confusing to you or seen as aggression and nonverbal actions.

There are at least five different aspects of OCD that can occur (*there are more, it is just I cannot get them out*). I am going to list them and they are going to sound very confusing, but when you read the explanations it will hopefully clear up the confusion. I will also add at the end each of aspect that can occur in OCD and autism, P.A.N.D.A.S, which is a medical condition and also co-exists with OCD.

They include:

1. OCD and rewinding
2. OCD and repeating
3. OCD and is this really happening
4. OCD and be quiet
5. OCD and thoughts

As an extra, I know that in the beginning of this book I explained and showed several scenarios in my life that actually show tactile cravings from autism. I will expand on it here as well with some more examples and how OCD *can* play into them as well.

OCD and rewinding:

When your child with autism is physically telling you to keep turning around and then telling you to go the other way (*or they are physically doing it because they are nonverbal which can be thought of as physical aggression*) and even doing this themselves, this can be a part of OCD. The person is thinking about how their body or another person's body was before and envisions it being wound up (*similar to a string*). They feel uncontrollable urges to get it back to being where it was in the first place (*similar to the string being let go of and unwinding to where it was in the beginning position*). It is very difficult because it can happen a lot and can take a very long time to happen because the mind is replaying and then has to start the replay in the mind over. It is a very upsetting part of the OCD and can cause both lots of frustration and anger, especially if they know what they are doing is irrational.

OCD and repeating:

I am pretty sure that you all are thinking of someone with autism and OCD touching things over and over (*yes this does happen because the mind has to say and feel that it "feels just right" until it can be finished*). However, in this case I am talking about autism and repeating verbal phrases. You may think this is just part of the echolalia, but it can extend to the autistic person wanting you to say a word repeatedly. This is because the brain has to keep on hearing the other person's voice (*reverberation of it*) until it sounds perfect and does not have mistakes in it like pronunciation, etc. As far as the other part, the touching things over and over does happen as well. It has to happen with the other person (*more often this is the person themselves with autism and OCD having to do it but in some cases extends to another member – meaning the autistic person is telling/making another person touch something over and over*) so they can see specific patterns and get the feeling of "feels just right." This again can be seen as the child controlling you, but it is actually the brain that is struggling to stop execution of repetitions and commands.

OCD and is this really happening:

Sometimes parents will see their child go into almost what seems like a schizophrenic state of mind (*which in autism is a part of the diagnosis included much of the time hence the word "spectrum," again it is schizotypal personality disorder, which also occurs in episodes of OCD*). The person with autism goes into a state like when someone (*even typical people have this happen but every once in a great, great while*) says, "Ok, I am really locking the doorknob now," or "Ok, I'm really going into the right entrance to the classroom now." It can lead to these things happening in the home or at school. The problem is it interferes with the person automatically just being in the regular clear state. For example, doing regular things like eating instead of thinking twice or walking into class instead of coming in late. This is a common thing in OCD for students to be late to class because they are having this type of OCD scenario.

OCD and be quiet:

Many people with OCD (*and like I said before, corresponds with autism*) have thoughts that go through their mind or repetitions of movies, words, etc. When another person starts to converse, they cannot handle it because it interrupts the process or the "playback" of what they are thinking. The person will then get angry and it can be thought of as sensory problems (*which at times it can be, but I am trying to say it can be this OCD scenario instead*) and can result in the person having to hold their ears shut and pace around the room. Some people even leave the room or get up (*in the middle of something*) so they can go and finish their quiet playback.

OCD and thoughts:

I am not going to go into this too much, but people with OCD can have irrational thoughts that scare them and also thoughts that can be very unpleasant or even mean. Many people would hear them and then get worried, but professionals and those who have this scenario with their OCD know that their thoughts are irrational and just in their brain; they wouldn't act upon the thoughts. Sometimes your child with autism/OCD will be upset and they may be having these different types of thoughts. Sometimes they may not be able to express them because of being nonverbal or communication problems overall. Many autistic individuals think that others think like them, so they have no clue it is not normal. (*As well as other things but this really is especially not understood until much later on*)

I am now going to talk about P.A.N.D.A.S, a medical disorder that co-exists with OCD as professionals are now finding. P.A.N.D.A.S is related to OCD by it getting worse with accompanying tics (*Tourette*) because of the part of the brain that gets affected and inflamed when many Strep infections (*as well as gamma*) are detected. You many notice that if your child already has autism and OCD (*even Tourette*), that when they get a strep infection they start to get worse in both their OCD and tic symptoms. Some children will have OCD with their autism and some tics, but the strep virus is causing the extreme degree of them as well as any clinginess/aggression. This is pediatric and for most pediatric doctors, the age they say pediatrics is, is to 21. They have even found this in adult patients who have gotten the strep infection because they started having tics/aggression/OCD tendencies, etc. They are still in research for this and the medicine used for it many times is Luvox, or its generic name, fluvoxamine. There are also subliminal therapy tracks they are putting together for downloading or getting on CDs for specific disorders. Ask your doctors, psychiatrists, and therapists about this type of treatment. The compositions use specialized technology and usually have a silent track so it is very convenient and not overwhelming. They actually work with the brainwaves, etc. One site I had gotten a track from is realsubliminal.com and found that it did help. Sometimes, as with any therapy, it can make things worse at first but they are really quite well created. An example would be you reading something in light gray on a page that you can barely see and yet your brain understands the message. Hard to comprehend I know, but that is another option since some of the medications can actually cause permanent issues such as Tardive dyskinesia. Also, the human body has strange reactions if it receives some of the medicine and has to go off of them for a little while. When your body is able to go back on the medication, it backfires and can create much worse problems.

I told you at the beginning of this chapter that I was going to add another example of how OCD can play into autism and tactile cravings (*although I gave several examples before and throughout*).

1. Many autistic people will rub their lips back and forth because it causes a strange tingling sensation, calming the body because it is receiving friction input. This can be related (*this specific one given here*) to autism and repetition.

2. The other thing that I want to mention that is very much OCD related directly and not as much autism unique related symptoms (*autism = its own unique symptoms but also a bunch of disorders in one*) is hair. It is when the person with autism

is getting sense cravings from ripping their hair out, many times more often than not their eyelashes. The technical term for this in OCD is trichotillomania. Fidgets can help with this and massaging the lashes in a circular motion with drops.

3. Seeing things in patterns is very OCD as well as stepping in counts. People with autism and dashes of OCD love to climb and jump. Climbing is a very common red flag of autism as well and is actually a talent. I would climb a ton and still climb (*well in rehab right now*) on rock walls, the rope walls, etc. Many autistic people will find things that are sturdy to put together in order to make a path to get to the top of something. This is a talent in autism known as visual and spatial awareness, not in the body but in objects. I would suggest also buying a safe walled trampoline for the intense sensory craving. It can help with the counting in OCD by counting jumps. For the climbing, it can help with them counting the rungs on the ropes to climb that you can purchase or use at parks.

When dealing with the OCD, you need to know that some compulsions when stopped, result in a terrible meltdown because the chemicals in the person's brain with OCD are flowing too fast. Beware of the situation and choose the battles.

I sincerely hope that this chapter has shed light into this deep-rooted condition related directly and also indirectly to autism so you can better understand the things that are going on with your child and why.

44

Autism and the Escape Artist

Many children who are lower functioning or who are just so occupied with getting things undone are escape artist autistics. This can be a horrifying thing for parents as it can put a damper on the home and even trips. Many times parents will have to put tons of locks and alarm systems in their houses, cameras, etc. so they can sleep through the night without worrying that their autistic child is going to escape once again. Then, they will have to change the locks or systems because the autistic person will figure out how to disarm the things the parents have worked so hard to install for safety purposes. Visually autistic individuals are usually very smart, so I guess if even for just a second you can kind of smile and think of your escape artist as a genius. To disarm all of those things is genius in my mind and considered by the police departments as remarkable; an amazing intellectual process as well as physical strength!

I have a couple suggestions that could help ease your worry and if not completely, at least you could sleep easier and trips could be a little more peaceful.

1. You can get a service dog – look at my chapter autism and service dogs in this book.
2. When you are using the safety installs, you can sort of use aversion therapy. For example, if there is something that he/

> she does not like to touch or even just go near by you could put that on or near the lock that you have set up. Then, when he/she goes there the next time to try and take everything apart and escape, they'll have an unpleasant experience for a moment. Hopefully, most of the time, they will not want to go back to it. Some parents fear that suggestion or are unsure about it. Alternatively, some are just not sure how to go about starting it. I say to them, "Just experiment to see what textures or things he/she recoils from. Also, this aversion is not like torturing him/her but it is basically a response to the brain that goes 'I don't want to touch/go near that, so I guess I can't get to those locks, etc.' You are not forcing these things upon them like sensory therapy."

I hope these suggestions have helped you and I pray for every family with autism who has a child who is an escape artist.

Before I go onto the next chapter, this certainly most likely would have gone into the service dogs chapter but I like to spread out my advice. My service dog could also track me back to where I need to go, as I can get lost easily. Therefore, she could help track me back with her nose and her special abilities. I had really wanted to mention that here because when a child escapes, you can have that attribute in a dog. I will now go on to chapter 45.

45

AUTISM AND "SPECIAL" TREATMENT

When people say, "I can't do that because it's not fair to give special treatment to him (*autistic child*) and not to these other children who are typical" it makes me cringe. I will explain why, in detail below.

Reversing the special treatment attitude:

I hear this quite a lot, but more from school districts like teachers, etc. My suggestion is this: (*ways to help look at it and maybe it will help them see it a different. way*)

*You will see why I use this symbol and why I say what I say when you continue to read on.

1. A person with diabetes needs "special" treatment so that their bodies are capable of being normalized. *You can use a different word for normal, because what is normal anyways?* A person would not say, "You can't have that snack right now because all the other kids are not allowed to have their snack before dinner," because it is a medical condition. AUTISM IS ALSO A MEDICAL CONDITION. It includes the way blood flows through the brain, the way the nerves are underneath the skin, the way the gut reacts to foods and how the system responds when it needs to use the bathroom, the way the balance control

> is in the brain and even the way the bones are positioned in some people with autism (*especially boys*), hence the odd gait. Therefore, he/she needs this treatment to be "normalized." It is not really special treatment, but treatment that NEEDS to be utilized to keep the person balanced like the other kids fortunately already have. Your autistic child then should have the same treatment - being able to be more "normal" in all ways. If the person is saying no "special treatment," then they are treating the other kids with special treatment and your autistic child is the one who is not receiving the same amount. These people are just looking at it from the opposite way. Since I am autistic, it is hard for me to write sometimes my thoughts, get them out right and say what I mean correctly.

I hope this can help you to get the services that your child both needs and deserves, meaning the treatment that normalizes the system so they have equal opportunity as the other children. If they are not getting equal opportunity, then they are not getting treated equally so the other kids are the ones getting treated special in a way. That is what I was trying to say above, only in a different light. Hopefully you can script this out for those you are trying to teach this message to by taking some specific examples from the "AUTISM IS A MEDICAL CONDITION" component and what I just said to sum this chapter up. That is the main message in a nutshell – the reversing the special treatment attitude.

46

AUTISM AND NEGATIVE TALK; IS THERE GOOD COMING OUT OF IT?

There have been many negative talks and comments about autism and autistic people on broadcasting networks, magazine articles and videos. That it is really sad and very unfortunate. However, I know this might seem way out of the galaxy (*LOL I like to use that term*), but could there be any good coming from all of this negative talk about autism?

The reason I say this is that look how much more autism is in the news and how much more autism is in the focus and spotlight now. Maybe this in a way will help people to say, "Ok, let's get serious and prove these people wrong (*even though we know they are wrong anyways*)." We are getting even more commercials out there to inform the public as well. Children's television shows are showing autism to children at younger ages so they grow up learning about it and accepting it (*and helping these people with autism*). I think autism will get more funding, more research, more services and more attention in the government, aside from what I said even above. Now, maybe that is just my "trying to stay positive" thinking, but is it maybe possible that these negative talks about autism *are* creating something good out of the bad? You decide. Either be angry (*of course, it is normal and understandable to a degree*) or be positive and when you are fighting back, fight back with forgiving affirmation and research. Maybe, just maybe, something good is coming out of these recent negative comments made about autism and autistic people.

47

Summary of My Book

I hope you have not only learned strategies, learned a little bit of how an autistic person views the world, but also had some fun while reading it. I feel *so blessed* to have been able to provide these things for others through my circumstances in my life. I feel like I was born to show things to others, and this was one of the ways.

God bless you all,

Kris

Now go out and raise Autism Awareness!

PS: I want to thank all those who inspired me to write this book, pushed me when I felt like it was impossible or when I was discouraged to write it and want to thank you all who read it; Above all I have to thank my Lord Jesus Christ. Without His mercy and grace, it would not have been possible. *My other message as an extra is dream big and do not give up. You can do it with help, faith and a strong foundation. If I can do this, you can do what you right now are thinking about and dreaming of, because I know when someone sees this message they are going to think of what they want in their life. Succeed and do it with love in your heart! If life leads you in another direction, it just means your calling is something you did not know was in your passion and strength. It might be several roads and you may have several callings, so go for them! Never give up on yourself.*

About the Author

Kristina DesJardins currently lives with her family in Gilbert, Arizona and enjoys visiting her relatives primarily in Minnesota, with one family living in California. She has friends in varying countries including Africa, France, Chile and a few others. She has many more aspiring dreams and enjoys training animals, creating over 20 forms of arts/crafts, playing several musical instruments, singing, learning languages and engaging in church activities. She's thankful to the people in her life that have crossed paths with her and thanks her Lord and Savior Jesus Christ for being able to have a full life. To her, all of her experiences have helped truly give her a full life; sometimes more than others may get to experience.

She dedicates this book to everyone who's trying to live their best life and those in her life who've passed away. She is also thankful to a number of directors and actors that she looked up to and helped her strive to keep her dreams alive, even if they don't know it. While there are many triumphs as well as challenges ahead as she battles various phases in life, she hopes to build upon those experiences to create movements. These movements are based on other experiences and situations that she's been a part of since she had typed this book in 2010. Big changes have already started, and she can't wait to see what becomes of them.

Stay tuned and keep in touch with her on her social media accounts. She says it strives her to do even more. If anyone is impacted by what she does then it is meaningful.

Thank you all for your part in following her life.

NEVER GIVE UP!

CPSIA information can be obtained
at www.ICGtesting.com
Printed in the USA
FSHW010956130122
87632FS

9 781956 110067